ENGINEERING DRAWING
WITH WORKED EXAM

F. Pickup, C.Eng., M.I.Prod.E.

and

M. A. Parker, T.Eng. (CEI), M.I.G. Tech.E.

Hutchinson of London

Hutchinson & Co (Publishers) Ltd
3 Fitzroy Square, London W1

London Melbourne Sydney Auckland
Wellington Johannesburg and agencies
throughout the world

First published 1960
Second impression 1961
Third impression 1962
Fourth impression 1963
Fifth impression 1964
Sixth impression 1966
Second edition, revised and metricated, 1970
Eighth impression 1971
Ninth impression 1972
Tenth impression 1973
Eleventh impression 1975
Third revised edition 1976
Thirteenth impression 1977
Fourteenth impression 1978

Printed in Great Britain by litho at The Anchor Press Ltd
and bound by Wm Brendon & Son Ltd
both of Tiptree, Essex

ISBN 0 09 126451 0

CONTENTS

PREFACE

The changes introduced in the 1972 revision of BS 308, Engineering Drawing Practice, have made a new edition of this book necessary. The general plan of the book, however, remains unchanged. The text has been kept to a minimum sufficient to outline the general principles of the subject, and worked examples have been freely used to enlarge on it. Each example shows the method of obtaining the solution, together with additional explanatory notes. For some topics where a solution on one drawing would have been difficult to understand the solution has been drawn in step-by-step form. The number of such solutions has been increased in this edition, and additional problems have also been provided.

The drawings have been completely redrawn and conform to the recommendations of BS 308: 1972. To mark the equal status given to First and Third Angle projection in this Standard, equal use has been made of the two systems. Chapters on conventions and technical sketching have been added, and other topics have been covered in more detail than previously. These include isometric and oblique projection, where the underlying principles of these projection systems have been explained more fully.

Several people have made suggestions for improvements in the book and have pointed out errors in previous editions. My thanks are due to them for their interest. I also acknowledge with thanks the permission given by the British Standards Institution for extracts from some of their Standards to be reprinted.

Hong Kong   M.A.P.

1976

1

LINES AND LETTERING

Types of line

The types of line for engineering drawings recommended by the British Standards Institution in BS 308:1972 are shown on page 4. Two line thicknesses are recommended: thick, 0·7 mm wide; and thin, 0·3 mm wide. These widths can be attained by using tubular ink pens, but for pencil drawings the recommendation can be interpreted as meaning that thick lines should be approximately twice as wide as thin lines.

The visible outlines of the object are drawn in continuous thick lines. They should be the most prominent lines on the drawing.

The hidden outlines of the object are represented by lines made up of short thin dashes. The dashes and the gaps between them must be consistent in length and approximately to the proportions shown on page 4. At corners and tangent points of arcs, dashes should meet.

The continuous thin line is used for dimension lines, projection lines, leaders for notes, hatching or section-lining, the outlines of adjacent parts and revolved sections, and fictitious outlines.

The limits of partial views and sections are shown by continuous irregular lines when the line is not an axis. These lines are thin and are drawn freehand.

Centre lines and the extreme positions of moveable parts are shown by thin chain lines. These comprise long dashes alternating with short dashes, not dots, proportioned approximately as shown on page 4. The lengths of the dashes and their spacing may be extended for very long lines.

Cutting planes for sections are represented by chain lines, thick at their ends and at changes of direction, thin elsewhere.

Thick chain lines indicate surfaces which have to meet special requirements. The lengths of the parts of these lines and the spacing between them should be similar to those of thin chain lines.

All chain lines must begin and end with a long dash. Centre lines should extend beyond the feature to which they refer for a short distance only, unless required for dimensioning. They should not be drawn through the spaces between views and must not end at another line of the drawing. Also they must cross each other at solid parts of the lines.

Chain lines having angles formed in them should be drawn with long dashes meeeting at the angles. Arcs should join at tangent points.

Arrowheads at the ends of dimension lines must touch the projection lines and those at the end of leaders must touch another line on the drawing. They must be sharp, black, filled-in and about 3 mm long.

Typical applications of the types of line are shown on page 5. For printing purposes all lines, except construction lines, must be black, dense and bold.

Pencil work

The best results are obtained, and the sharpening of pencils is reduced, if all straight lines are drawn with chisel-edged pencils, and lettering, arrowheads and continuous irregular lines are done with conical pointed pencils. Suitable grades of pencil for use on cartridge paper are HB or H for outlines, lettering and arrowheads, H for all thin lines, and 2H or 3H for construction lines. For detail paper, pencils should be a little harder and for linen a little softer. The more abrasive the paper the harder the pencils should be. Bold, black, dense lines can only be produced by sharp pencils, and pencil points should be frequently sharpened on an old smooth file or a glass paper block.

Compass work

Compass leads should be sharpened by rubbing one side of them down on a file or glass paper block until a sharp, curved edge is produced. Chisel or conical points should not be used as they tend to produce lines of variable thickness. Lines drawn with a compass tend to be less bold and black than those drawn with a pencil as less pressure can be applied. To compensate for this a softer lead should be used in compasses.

Compasses should be fitted with a shouldered needle point to prevent a large hole being made in the paper at the centre of concentric circles. This shouldered needle should project a little way beyond the lead when the compasses are closed, so that the needle pierces the paper before the lead makes contact. This will prevent the needle slipping as the circle is being drawn. Small circles can only be drawn successfully if the compass point is sharpened properly and correctly adjusted to the needle point.

Radius templates

When many small radii have to be drawn, for example on drawings of castings and forgings, it is more convenient and much quicker to use radius templates rather than compasses. A conical pointed pencil, of the same grade as that for outlines, should be used.

Lettering

The essential features of lettering on engineering drawings are legibility, uniformity and the ability to be produced rapidly. Legibility and speed are achieved by the use of a block, single-stroke style which may be either upright or sloping. Students are recommended to use the upright style as it is easier to produce. Single-stroke lettering has all the strokes of uniform thickness, and each stroke is produced by one movement of the pencil. Capital letters are preferred to lower case ones, being less congested and less likely to be misread when reduced in size on prints. Lower-case letters should, however, be used when they are part of a standard symbol, code or abbreviation.

A suitable alphabet and figures are shown on page 6 and this model should be consulted frequently in the early stages until the character forms are memorized. Note that the characters have the simplest possible forms. Flourishes and ornament are out of place on an engineering drawing.

All pencil lettering should be produced freehand and drawn between a pair of faint guide lines. For dimensions and notes a character height of about 3 mm should be used, and characters should be about the same width. Titles are generally made in larger characters. Characters must touch the guide lines and be consistent in width. As an aid to spacing words consistently, imagine an 'I' to be placed between them. The space between lines of lettering should not be less than half the character height.

As an aid to reading them, all notes should be lettered to read from the bottom of the drawing. Notes should not be underlined. If a note is important and needs to be emphasized, larger characters should be used.

The decimal marker used with metric units is a point which should be bold, given a full letter space, and be placed on the base line. It is also recommended that where there are more than four figures to the right or left of the decimal marker, a full letter space should be left between each group of three figures, counting from the decimal marker. Dimensions which are less than unity should be preceded by the cipher 'O'.

These points are illustrated on page 6.

Scales

All drawings should be made full size if possible, but if the size of the object is such as to make this impossible they must be drawn in proportion, that is, to a uniform scale. The scale used must be stated on the drawing as a ratio or representative fraction, for example scale 1 :2, which means half full size. It is common for a note to warn against scaling the drawing, since prints may stretch or shrink.

Drawings are sometimes printed in an enlarged or reduced form. In such cases it is useful for the scale to which the drawing has been pro-

LINES AND LETTERING

A ————————————————— Visible outlines and edges
  CONTINUOUS (THICK)

B ———————————————— Hidden outlines and edges
  SHORT DASHES (THIN)

C ————————————————— Dimension and leader lines
  CONTINUOUS (THIN)   Hatching
                                                     Outlines of adjacent parts
                                                     Outlines of revolved sections
                                                     Fictitious outlines and edges

D ～～～～～～～～～～～ Limits of partial views or
  CONTINUOUS IRREGULAR (THIN)   sections when the line is not
                                                     an axis

E ————————————————— Centre lines
  CHAIN (THIN)   Extreme positions of moveable
                                                     parts

F ————————————————— Cutting planes
  CHAIN (THIN, ENDS AND CHANGES OF
  DIRECTION THICK)

G ————————————————— Indication of surfaces which
  CHAIN (THICK)   have to meet special requirements

TYPES OF LINE FROM BS 308 : PART I : 1972

ARROWHEADS — Sharp , black and filled — in. About 3mm long.

4

LINES AND LETTERING

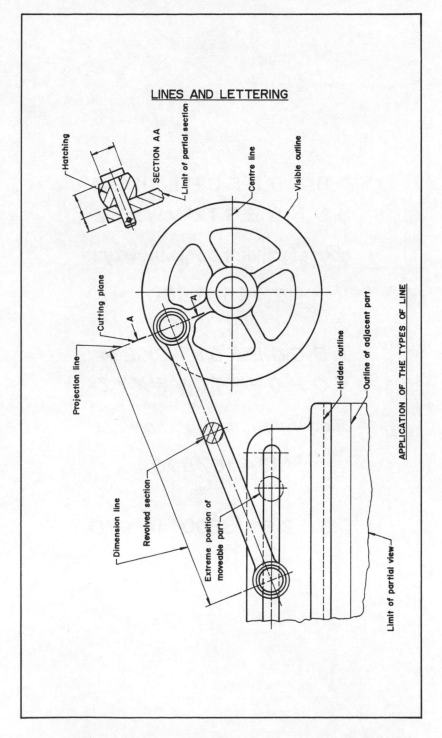

Hatching

Limit of partial section

SECTION AA

Centre line

Visible outline

Cutting plane

Projection line

Dimension line

Revolved section

Extreme position of
moveable part

Hidden outline

Outline of adjacent part

Limit of partial view

APPLICATION OF THE TYPES OF LINE

A B C D E F G H I J K L M
N O P Q R S T U V W X Y Z

abcdefghijklmnopqrstuvwxyz

1234567890

A B C D E F G H I J K L M
N O P Q R S T U V W X Y Z

abcdefghijklmnopqrstuvwxyz

1234567890

0.5   2.6   3800   14 970

duced to be drawn along the margin of the original sheet.

When components are drawn larger than full size it may be useful to show an undimensioned full-size pictorial or orthographic view. However, this may be misleading if the drawing is reproduced at a ratio other than 1:1.

Scale multipliers and divisors of 2, 5 and 10 are recommended and the representative fractions of the most commonly used scales are

1:1	Full size
1:2	Half full size
1:5	One-fifth full size
1:10	One-tenth full size
2:1	Twice full size
5:1	Five times full size
10:1	Ten times full size

LINES AND LETTERING PROBLEMS

1 Copy Examples 1 to 8 on page 8 with dimensions, using the lines indicated. Make all lines, except construction lines, black, bold and dense. Observe the distinction between thick and thin lines and keep line thicknesses consistent throughout.

2 Letter the alphabet in capital letters and the figures up to 9 in 3 mm, 6 mm and 10 mm characters. Use faint guide lines and keep character widths and spacing consistent.

3 Letter the following dimensions in 3 mm, 6 mm and 10 mm figures.
0.75   3.16   1.65   442   1 290   32 780

4 Letter the notes as set out below in 3 mm characters. Use faint guide lines and leave 3 mm between lines of lettering.

(a) 4 HOLES ϕ5.5
SPACED AS SHOWN

(b) 3 BOSSES ϕ16
EQUISPACED ON 108
PCD

(c) DRILL AND REAM
IN POSITION FOR
ϕ4.7 TAPER PIN

(d) 24 SERRATIONS 30 LG
ONE LEFT UNCUT
WHERE SHOWN

(e) C'BORE BRONZE BUSH
ϕ24 × 9.5 DEEP
ON ASSEMBLY

(f) CADMIUM PLATE 0.05
THICK ALL OVER
EXCEPT WHERE
MARKED xxxx

LINE PROBLEMS

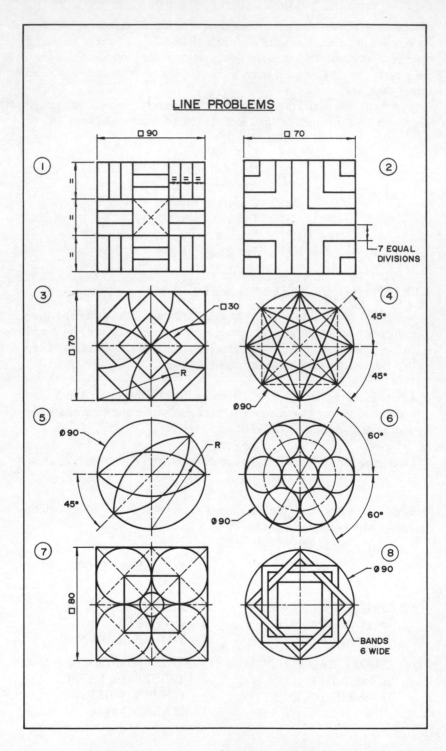

8

2

GEOMETRICAL CONSTRUCTIONS

To bisect a line (Figure 1)
Draw the given line AB. With centres A and B and radius R greater than half of AB, draw arcs to intersect at C and D. Join CD, when E will be the mid point of the line. Also CD will be perpendicular to AB.

To divide a line into a number of equal parts (Figure 2)
From one end of the given line (say A), draw AC at any convenient angle. Using dividers or a scale, mark off from A on AC the required number of equal parts, making them of any suitable length. Join the last point to B on the given line, and through the other points draw parallels to this line to cut the given line. This construction makes use of the properties of similar triangles.

To divide a line in a given proportion (Figure 3)
Suppose the proportion to be 2:3. Using the previous construction, proceed as if to divide the line into 5 parts (2 plus 3) but only draw a line through point 2 on AD. Then AB will be divided in the required proportion.

To bisect an angle (Figure 4)
Draw the given angle ABC and from the apex B draw an arc of radius R to cut AB and CB at D and E. R may be any convenient radius. With D and E as centres and radius R^1, draw two arcs to meet at F. Again, R^1 may be any convenient radius. Join FB to bisect the angle.

To find the centre of an arc (Figure 5)
Select three points, A, B and C on the arc and join AB and BC. Bisect these lines and produce the bisectors to meet at O. O is the centre of the arc.

To inscribe a circle in a triangle (Figure 6)
Draw the given triangle ABC and bisect any two angles. Produce the bisectors to intersect at O, which is the centre of the inscribed circle.

GEOMETRICAL CONSTRUCTIONS

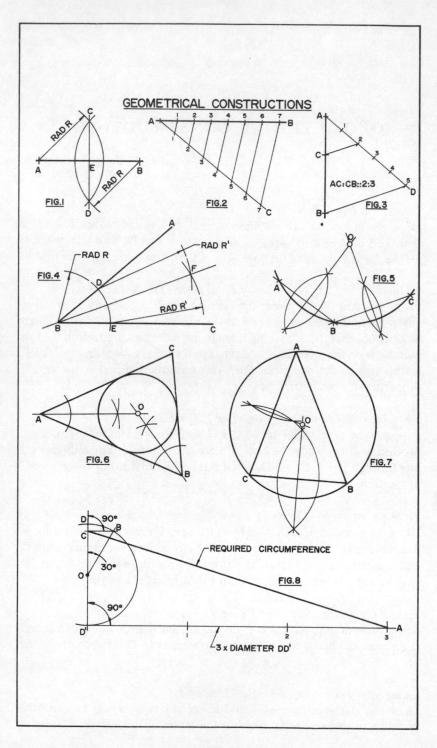

FIG.1

FIG.2

AC:CB::2:3

FIG.3

RAD R

RAD R'

RAD R'

FIG.4

FIG.5

FIG.6

FIG.7

REQUIRED CIRCUMFERENCE

FIG.8

90°

30°

90°

3 × DIAMETER DD'

To draw the circumscribing circle of a triangle (Figure 7)
Draw the given triangle ABC. Bisect any two of the sides and produce the bisectors to intersect at O. O is the centre of the circumscribing circle.

To find graphically the circumference of a circle (Figure 8)
Draw a semicircle with diameter DD' equal to the diameter of the given circle. From D' draw D'A at right angles to DD' and mark off along it from D' three diameters, thus finding point 3. From the centre O of the semicircle draw OB at 30° to OD. From B draw BC at right angles to OD. The required circumference is the line C–3. This construction will be found useful in some loci and development work.

To draw a regular hexagon given the distance across flats (Figure 9)
Draw a circle having a diameter equal to the distance across flats. Draw tangents to this circle with a 60° set square to produce the hexagon.

To draw a regular hexagon given the distance across corners (Figure 10)
Draw a circle having a diameter equal to the distance across corners and step off the radius round it to give six equally spaced points. Join these points to form the hexagon.

To draw a regular octagon given the distance across flats (Figure 11)
Draw a circle with a diameter equal to the distance across flats. Construct the octagon by drawing tangents at 45° to this circle.

To draw a regular octagon given the distance across corners (Figure 12)
Draw a circle with a diameter equal to the distance across corners. Across this circle draw vertical and horizontal diameters and two at 45°. Join the eight points so obtained on the circle to form the octagon.

To draw a regular pentagon given the length of the side (Figure 13)
Draw the given side AB and with centres A and B draw two circles of radius AB to intersect at C and D. Join CD. With centre D and radius AB draw an arc to cut the previously drawn circles at E and F and CD at G. Join EG and FG and produce to H and J on the first two circles. These points are corners of the pentagon. With centres H and J and radius AB draw arcs to meet at K. Join AJKHB to complete the pentagon.

To draw any regular polygon given the length of the side (Figure 14)
Suppose the polygon to have seven sides. Draw the given side AB and on it as base construct two triangles with base angles of 45° and 60°. The apices of these triangles, marked 4 and 6 in the figure, are respectively

11

GEOMETRICAL CONSTRUCTIONS

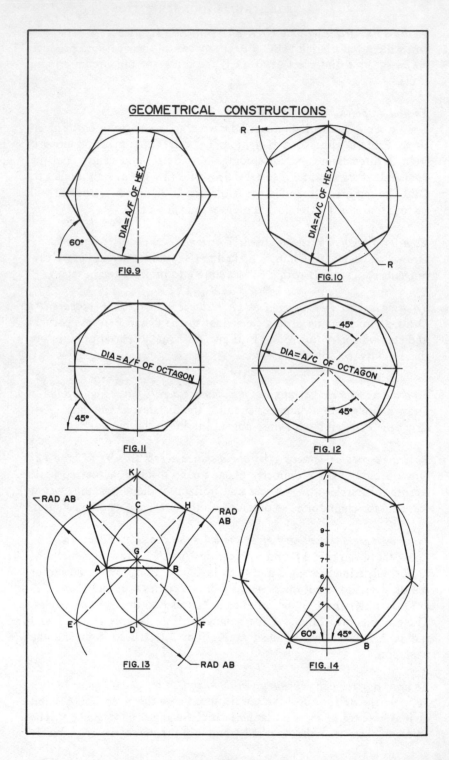

FIG.9

FIG.10

FIG.11

FIG.12

FIG.13

FIG.14

12

the centres for circumscribing circles of regular polygons with four and six sides of length AB. Bisect the line 4–6 and obtain point 5. This is the centre of the circumscribing circle for a regular pentagon of side AB. Along 4–6 produced step off length 4–5 to obtain point 7. This is the centre of the circumscribing circle for a regular heptagon of side AB. Draw this circle with radius A–7, and step AB round it six times. Join the points so obtained to give the required polygon.

If length 4–5 is stepped off from point 7 as many times as necessary, the centres for circumscribing circles of regular polygons with any number of sides of length AB may be found.

GEOMETRICAL CONSTRUCTION PROBLEMS

In these problems the solution must be lettered and dimensioned as stated in the question. The lines of the solution must be black with the construction lines faint.

1 Draw a line AB 80 mm long and bisect it.

2 A spindle is shown in Figure 1 in which the lengths of the various diameters are expressed as fractions of the total length. Copy the drawing obtaining the lengths by construction.

3 Draw a line AB 165 mm long and divide it in the proportion 3:4:2.

4 Three points, X, Y and Z are shown in Figure 2 positioned relative to two axes OA and OB. Draw the Figure and draw an arc to pass through the three points.

5 Using the angles of the 45° and 60° set squares as a basis, construct the following angles by bisection:
   (a) $22\frac{1}{2}°$ (b) 15° (c) $52\frac{1}{2}°$ (d) $112\frac{1}{2}°$ (e) $37\frac{1}{2}°$ (f) $146\frac{1}{4}°$.

6 A triangle ABC stands on side AB as base and has the following dimensions: AB 89 mm, AC 76 mm, angle CAB $67\frac{1}{2}°$. Construct the triangle and draw the inscribed circle.

7 Construct a triangle ABC on AB as base with AB 70 mm, AC 57 mm, BC 76 mm and draw the circumscribing circle.

8 Find graphically the circumference of a circle of diameter 70 mm, and check the result by calculation.

9 Construct regular hexagons to the following dimensions:
   (a) 90 mm across flats (b) 95 mm across corners.

10 Draw two regular octagons, one 76 mm across flats and the other 82 mm across corners.

13

11 Construct a regular pentagon with sides 32 mm long.

12 Draw a regular heptagon with sides 38 mm long.

13 A view of a drilling template is shown in Figure 3. Copy this view full size constructing the centres for the 6 mm diameter and 8 mm diameter holes.

GEOMETRICAL CONSTRUCTION PROBLEMS

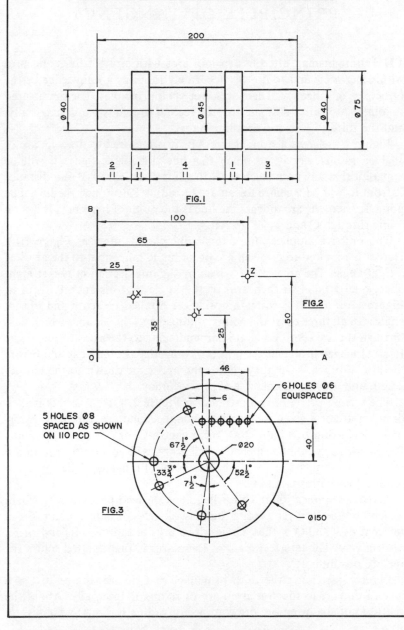

FIG.1

FIG.2

FIG.3

5 HOLES Ø8 SPACED AS SHOWN ON 110 PCD

6 HOLES Ø6 EQUISPACED

15

3

PRINCIPLES OF TANGENCY

THE draughtsman often has to join arcs with straight lines and arcs with other arcs, and to do this accurately requires a knowledge of the principles of tangency. There are three such principles. The first is used to join an arc with a straight line, the second to join two arcs externally, and the third to join two arcs internally.

Figure 1(a) shows a straight line AB. It is required to draw an arc of a given radius, say R, to touch this line. From the figure it will be apparent that any point such as C, in a line parallel to AB and distance R from it, will be a centre for an arc of radius R to touch the line. The point T where the arc touches the line is the point of tangency. It lies on a line through C and at 90° to AB.

The practical application of this principle is shown in Figure 1(b). Here it is required to draw an arc of radius R tangential to the arms of a right angle. The arc centre is given by the intersection of two straight lines, one parallel to each arm of the angle and distance R from it. Figures 1(c) and 1(d) show the principle applied to acute and obtuse angles. In all three cases the points of tangency T lie on lines which pass through the arc centres and are perpendicular to the arms of the angle. In all tangency problems the points of tangency should be found before the arc is drawn, since they enable the arc to be drawn to the correct length and are thus a valuable aid to accuracy.

The second principle is illustrated in Figure 2. Here it is required to draw an arc of a given radius r to touch a second arc, radius R, externally. Referring to Figure 2(a), from centre O draw an arc of radius R + r. Then any point such as C, on this arc, will be a centre for an arc of radius r to touch the given arc. The point of tangency T lies on a line joining the centres O and C.

A typical application of this principle is shown in Figure 2(b), where a blend radius R_1 is required to touch two circles of radii R and r. From the centres P and Q of these circles draw arcs of radii $R + R_1$ and $r + R_1$ respectively. The intersection O of these arcs is the required centre for the arc of radius R_1.

Figure 3(a) shows the third principle, that of drawing an arc of a given radius r, to touch a given arc of radius R internally. About the centre O of the given arc draw an arc of radius R − r. Any point C on

16

PRINCIPLES OF TANGENCY

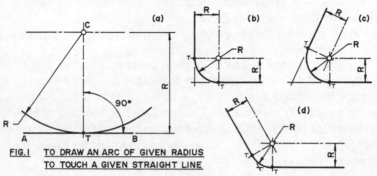

FIG.I   TO DRAW AN ARC OF GIVEN RADIUS
         TO TOUCH A GIVEN STRAIGHT LINE

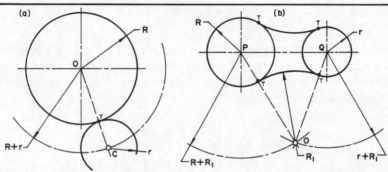

FIG.2   TO DRAW AN ARC OF GIVEN RADIUS TO TOUCH A GIVEN ARC EXTERNALLY

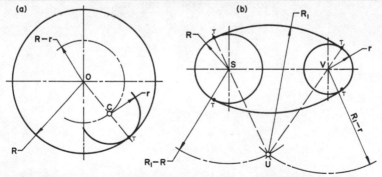

FIG.3   TO DRAW AN ARC OF GIVEN RADIUS TO TOUCH A GIVEN ARC INTERNALLY

this arc will be the centre for an arc of radius r to touch the given arc internally. As in Figure 2, the point of tangency T lies on a line joining the centres O and C.

In the practical application shown in Figure 3(b), the two given circles are to touch the blend radius R_1 internally. To find the centre of the blend radius draw arcs of radii $R_1 - R$ and $R_1 - r$ about the centres S and V respectively of the given circles. Where these arcs intersect at U is the centre for the blend radius R_1.

It should be noted that in all problems in tangency, the intersection of two lines is always required to find the centre of an arc.

When drawing arcs tangential to each other it is best to draw complete faint circles first, and when the drawing is finished, to darken the part required. When arcs have to be drawn tangentially to a straight line, the best results are obtained if the straight line is darkened after the arcs have been completed. It is easier to compensate for small errors by moving a straight line slightly, than by trying to adjust the position of an arc centre.

TANGENCY PROBLEMS

In the following problems, construction lines must not be erased and all centre lines must be shown as in the figures. For lettering and dimensioning practice some or all of the problems may be dimensioned. Suitable scales are

> Problems 1 to 9 full size
> Problems 10 and 11 half full size
> Problems 12 and 13 full size
> Problem 14 half full size
> Problems 15 and 16 full size

TANGENCY PROBLEMS

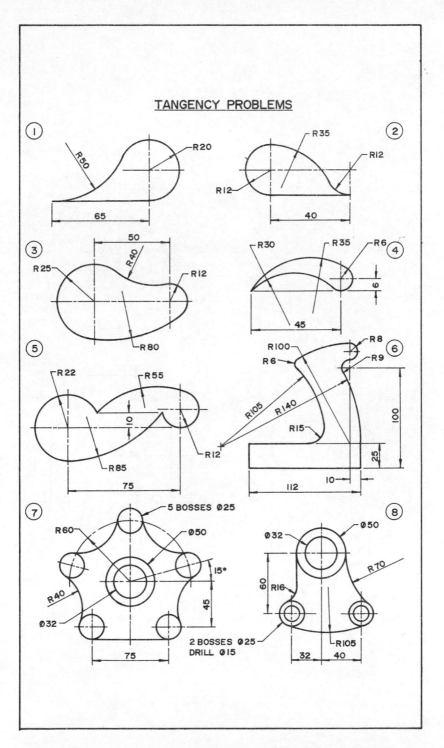

TANGENCY PROBLEMS

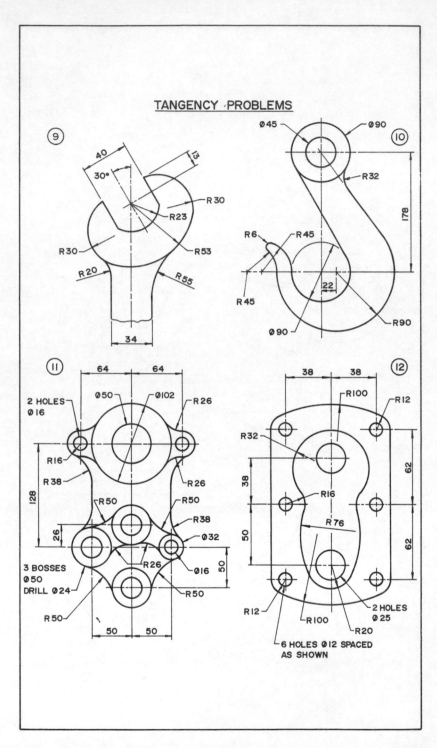

20

TANGENCY PROBLEMS

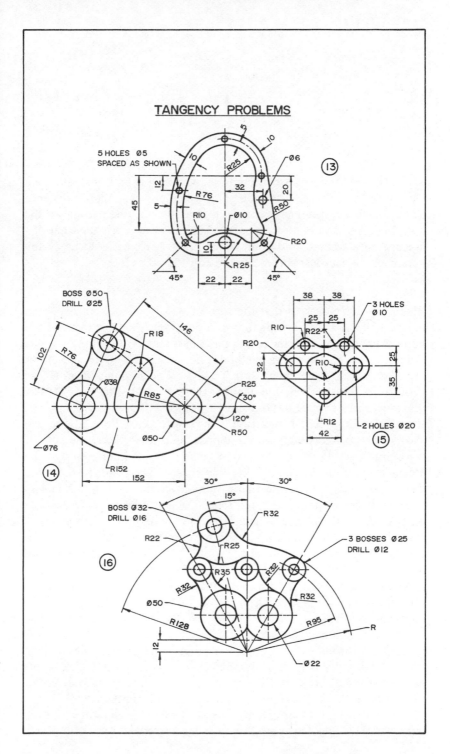

4

LOCI

WHEN a point moves according to a given law its path is said to be a locus. For example, the locus of a point which moves so that its distance from a fixed straight line is constant, is a straight line parallel to the fixed line. Again, a point moving so that it is always at a constant distance from a second fixed point traces out a circle as its locus. Many important geometrical curves may be considered as loci.

Ellipse
When a point moves so that the sum of its distances from two fixed points, called focal points or foci, is a constant, then the locus of the point is an ellipse. The constant is the major axis of the ellipse.

To draw an ellipse by this method, first set out the major and minor axes AB and CD as in Figure 1(a). To find the focal points draw an arc with radius equal to the semi-major axis AO, centred at C or D. The points F_1 and F_2 where this arc cuts the major axis are the focal points, since $CF_1 + CF_2 = AB$, thus satisfying the law for the ellipse. Figure 1(b) shows how to find points on the ellipse such as P. Arcs, the sum of whose radii is AB, are drawn from F_1 and F_2. The intersection of the arcs is a point on the curve. Note that three other points in the other three quadrants of the ellipse can be found using the same radii X and Y.

The number of points required to ensure an accurate curve depends on the size of the ellipse, but in any case a minimum of four points should be determined in each quadrant. Join the points with a french curve, not freehand, noting that the ellipse consists of four identical quadrants and that it crosses the axes at right angles.

An ellipse may also be drawn by the concentric or auxiliary circles method, shown in Figure 2. Draw the axes AB and CD and draw circles on them as diameters. These circles are the auxiliary circles of the ellipse. Divide the circles into a number of parts, preferably equal, by radial lines through O. Through the points where these lines cut the major auxiliary circle drop perpendiculars, and through the points where the radial lines cut the minor auxiliary circle draw horizontals to cut the verticals. These intersections are points on the ellipse.

In Figure 2(b), for clarity, only two points have been plotted in each quadrant, but, as noted above, for an accurate curve at least four points

LOCI

ELLIPSE CONSTRUCTIONS

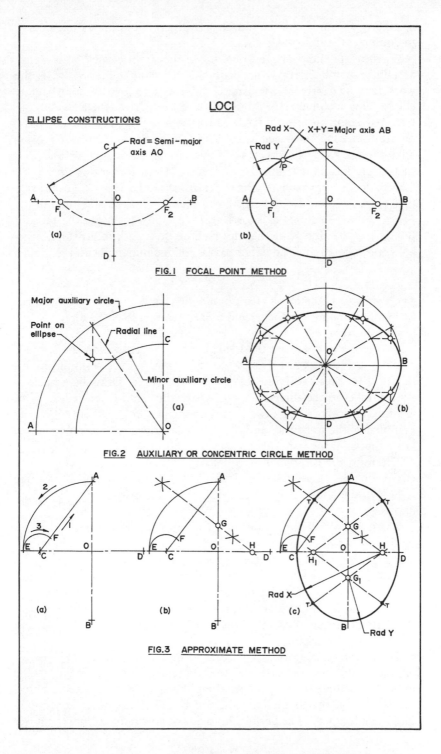

FIG.1 FOCAL POINT METHOD

FIG.2 AUXILIARY OR CONCENTRIC CIRCLE METHOD

FIG.3 APPROXIMATE METHOD

23

in each quadrant should be found. Join the points with a french curve to complete the drawing.

When many ellipses have to be drawn the simplest method is to use an ellipse template. However, if no suitable template is available, the work may be shortened by using circular arcs to give an approximate ellipse. One such method is shown in Figure 3. Draw the axes AB and CD intersecting at O. Join CA and with centre O and radius OA draw an arc to cut DC produced at E. With C as centre and radius CE draw an arc to cut AC at F. Bisect AF. Where the bisector cuts AB and CD, or CD produced, at G and H, are centres for arcs, radii X and Y, to form half the approximate ellipse. Transfer G and H to the other side of O to give G_1 and H_1, the centres for the radii for the other half of the figure. Join the centres as shown in Figure 3(c) to locate the tangent points T of the arcs. Note that this method does not produce an acceptable approximation if the minor axis is small compared with the major.

Parabola

The parabola is the locus of a point which moves so that its distances from a fixed point, the focus, and a fixed straight line, the directrix, are always equal.

To draw the curve by the locus method, first position the focus F and the directrix as in Figure 4(a). The vertex V may be found by bisecting FG, thus satisfying the law for the curve. Draw lines such as AA, parallel to the directrix, and by drawing an arc centred at F, of radius GH, to cut AA, fix two points on the parabola. Other points may be found in the same way and the parabola drawn with a french curve. Note that at V the parabola crosses the axis at 90°.

The ratio PF:RP, which is equal to unity, is the eccentricity of the curve. For the ellipse the eccentricity is less than unity and if its value is known the ellipse may be drawn in a similar way to that described above.

When the overall dimensions of the parabola are known the curve may be drawn by the circumscribing rectangle method shown in Figure 5. Use the span and rise to lay out a rectangle and divide it in two by the axis of symmetry. Divide the sides of the two rectangles so obtained into the same number of equal divisions. Draw lines parallel to the axis through one set of divisions and join each of the other sets to O. Where the parallel through 2 cuts the line O to 2 is a point on the parabola, and so on.

The circumscribing rectangle method may also be used to draw an ellipse. In this case the sides of the rectangle are equal in length to the axes of the ellipse. The divisions along the span in the parabola above are laid out along the major axis, and the divisions on the rise along half the minor axis. The parallel lines above now radiate from the lower

LOCI

PARABOLA CONSTRUCTIONS

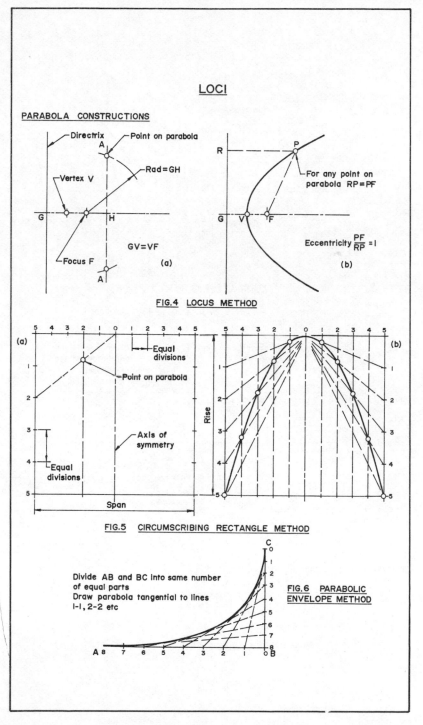

FIG.4 LOCUS METHOD

FIG.5 CIRCUMSCRIBING RECTANGLE METHOD

Divide AB and BC into same number
of equal parts
Draw parabola tangential to lines
1-1, 2-2 etc

FIG.6 PARABOLIC
ENVELOPE METHOD

LOCI

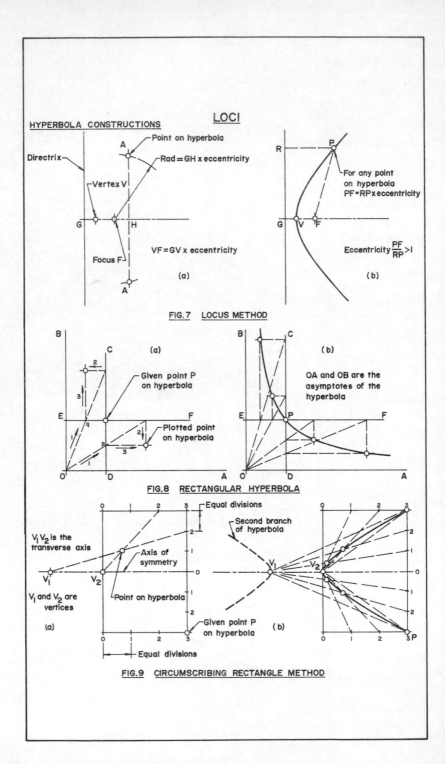

HYPERBOLA CONSTRUCTIONS

Directrix

Point on hyperbola

A

Rad = GH x eccentricity

Vertex V

G ——— H

Focus F

VF = GV x eccentricity

(a)

R ——— P

For any point
on hyperbola
PF = RP x eccentricity

G V F

Eccentricity $\dfrac{PF}{RP} > 1$

(b)

FIG.7 LOCUS METHOD

B

C

(a)

2

3

Given point P
on hyperbola

E ——— F
q

Plotted point
on hyperbola

2

3

1

O D A

B

C

(b)

OA and OB are the
asymptotes of the
hyperbola

E P F

O D A

FIG.8 RECTANGULAR HYPERBOLA

Equal divisions

0 1 2 3

2

$V_1 V_2$ is the
transverse axis

1

Axis of
symmetry

0

V_1 V_2

V_1 and V_2 are
vertices

1

Point on hyperbola

2

(a)

Given point P
on hyperbola

0 1 2 3

Equal divisions

Second branch
of hyperbola

0 1 2 3

2

1

V_1 V_2 0

1

2

(b)

0 1 2 3 P

FIG.9 CIRCUMSCRIBING RECTANGLE METHOD

end of the minor axis and the points are plotted as before.

Figure 6 shows the parabolic envelope method. The lines AB and BC are drawn at right angles and divided as shown into the same number of equal parts. Corresponding points on each line are joined and the curve is drawn tangentially to these straight lines. This method can be used whatever the angle between AB and BC.

The parabola has the property that if a source of light, sound or heat is placed at the focus, rays reflected from the parabola form a parallel beam. This property is made use of in hand torches, car headlamps and some electric fires. Conversely, rays falling on the parabola from outside will be reflected to the focus. This explains the shape of some radar and radio dishes which have the form of a paraboloid. This is the surface generated by revolving a parabola about its axis.

Hyperbola

The hyperbola is the locus of a point which moves so that the ratio of its distances from the focus and directrix is constant and greater than 1. The eccentricity is thus greater than 1.

To draw the curve for a given ratio, say 3/2, first position the focus and directrix as shown in Figure 7. The vertex is found by dividing GF into five (3 plus 2) equal parts, when V will be two divisions from G. To fix further points, draw any line such as AA, parallel to the directrix, and with centre F and radius GH times the eccentricity, draw arcs to cut AA above and below the axis. These points will be on the curve. The distances GH should be so chosen that the radius from the focus is easily calculated.

Lines which are tangents to the hyperbola at infinity are called asymptotes. When these are given, together with one point on the curve, the method shown in Figure 8 may be used to construct the hyperbola. Draw the asymptotes OA and OB and position the given point P. Through P draw CD and EF parallel to the asymptotes. Through any points such as p and q on CD and EF draw lines marked 1, 2 and 3 to plot further points on the curve.

When the asymptotes are at right angles as in Figure 8, the hyperbola is called rectangular or equilateral. The method may however be used whatever the angle between the asymptotes, provided that CD and EF are each drawn parallel to one asymptote.

The hyperbola has two branches, each of which has a focus and directrix. Both branches have the same eccentricity. When both vertices are known, together with a point on one branch, both branches may be drawn by the circumscribing rectangle method shown in Figure 9.

Position the vertices V_1 and V_2 and the given point P, and construct a rectangle through P and its associated vertex. Divide the rectangle in two by the axis of symmetry. Divide the sides of the two rectangles so

LOCI

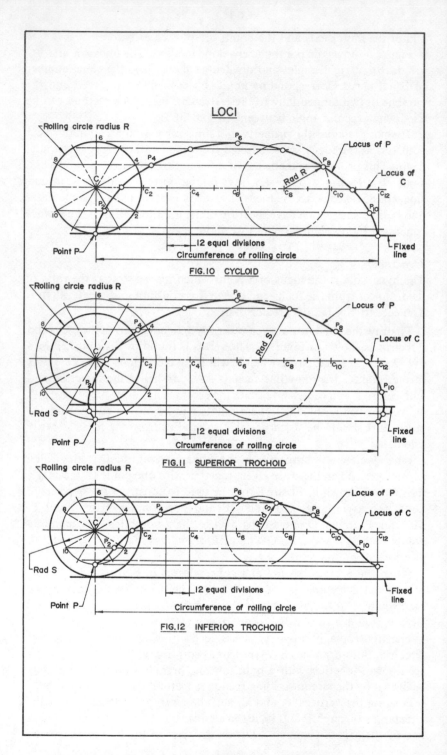

FIG.10 CYCLOID

FIG.11 SUPERIOR TROCHOID

FIG.12 INFERIOR TROCHOID

obtained into the same number of equal parts and join the vertices to them as shown, to plot points on one branch of the hyperbola. The second branch may be drawn in the same way or by making use of the symmetry of the curves.

The ellipse, parabola and hyperbola are conic sections. That is, they may be produced by passing planes through a right cone. This aspect is treated in Book 2.

Cycloid

This curve is the locus of a point on the circumference of a circle which rolls without slipping along a fixed straight line.

To construct the curve, refer to Figure 10. Draw the rolling circle and fixed straight line and fix the initial position of the point P which is to trace out the locus. Draw the locus of C, the centre of the rolling circle. This will be a straight line, equal in length to the circumference of the rolling circle, parallel to the fixed line, and the radius R of the rolling circle from it. Divide the locus of C and the rolling circle circumference into the same number of equal parts – twelve is a convenient number. Through the points on the rolling circle circumference draw parallels to the fixed line. The points on the locus of C represent twelve successive positions of the rolling circle centre. When the centre has moved to the first of these positions, P has risen to the level of the line through 1 on the rolling circle. Its position on this line is found by cutting the line with an arc equal to the rolling circle radius, struck from the new centre. Other points are found in a similar manner. The locus may be continued indefinitely, but repeats itself for each complete revolution of the rolling circle.

Superior and inferior trochoids

When the point which traces the locus is outside the rolling circle the locus produced is a superior trochoid, shown in Figure 11. When it is inside, the locus is an inferior trochoid, shown in Figure 12. The construction for both is very similar to that for the cycloid. It should be noted however, that in each case the circumference of the rolling circle is laid out along the locus of its centre C, and the circle through P is divided into twelve equal parts, not the reverse.

Epicycloid and hypocycloid

When a circle rolls without slipping round the outside of a fixed circle (called the base circle), a point on the circumference of the rolling circle traces out an epicycloid. When the circle rolls round the inside of the base circle the resulting locus is a hypocycloid.

The construction for both curves is again very similar to that for the cycloid. Now, however, use must be made of the angle θ in Figure 13,

LOCI

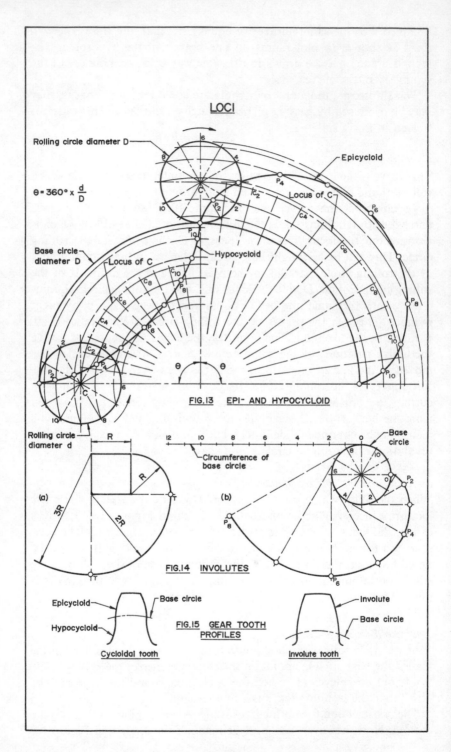

Rolling circle diameter D

$\theta = 360° \times \dfrac{d}{D}$

Epicycloid

Locus of C

Base circle diameter D

Locus of C

Hypocycloid

FIG.13 EPI- AND HYPOCYCLOID

Rolling circle diameter d

Circumference of base circle

Base circle

(a)

(b)

FIG.14 INVOLUTES

Epicycloid — Base circle

Hypocycloid —

FIG.15 GEAR TOOTH PROFILES

Involute

Base circle

Cycloidal tooth

Involute tooth

to obtain the twelve positions of the centre of the rolling circle. θ is the angle subtended at the centre of the base circle by an arc equal to the circumference of the rolling circle. This arc will be passed over by the rolling circle in one revolution, and will have a length of πd. The ratio of this arc to the circumference of the base circle will be the same as the ratio of θ to 360°.

$$\text{Thus} \quad \frac{\theta°}{360°} = \frac{\pi d}{\pi D}$$

$$\text{and} \quad \theta° = 360° \times \frac{d}{D}$$

The base circle, rolling circle and θ can now be laid out and the rolling circle and θ divided into the same number of equal parts, twelve for convenience. The remainder of the construction should be followed easily from the figure. Students should note, however, that the arc through points 3 and 9 is not the path of C, the rolling circle centre.

Involutes

If a straight line is rolled round a polygon without slipping, points on the line will trace out involutes.

Figure 14(a) shows the construction for the involute of a square of side R. When the line rolls round the square it will pivot on successive corners, so the locus consists of a series of circular arcs whose radius increases by R as each corner becomes the pivot point.

The circle may be thought of as a polygon with an infinite number of sides, and Figure 14(b) shows the construction for its involute. First draw the base circle and in some convenient position a line equal in length to its circumference. Divide both into the same number of equal parts (say twelve). From the points on the circle draw tangents as shown, to represent successive positions of the generating line. When this line is in the position of the tangent at point 1 it will have rolled over one-twelfth of the circumference. Take one-twelfth from the straight line along which the base circle circumference was set out, and mark it off along the tangent at point 1, thus obtaining a point on the curve. Repeat the procedure stepping off two-twelfths, three-twelfths of the circumference on succeeding tangents.

The cycloid, epicycloid, hypocycloid and involute have an important application in the profiles of gear teeth, as shown in Figure 15. In the cycloidal system the profile consists of parts of an epicycloid and hypocycloid, above and below the base circle, whilst the rack tooth profile is part of a cycloid. In the involute system the part of the profile above the base circle is an involute and involute rack teeth have straight profiles. Involute gears are dealt with more fully in Book 2.

31

Helices

A helix is the locus of a point which moves round the circumference of a right cylinder, at the same time moving axially, the ratio of the two movements being constant. The axial movement during one revolution is called the lead. Helices may be right- or left-handed. If the point moves clockwise and away from the observer when the cylinder is viewed axially, the helix is right-handed. If the point moves anticlockwise and away from the observer, the helix is left-handed. The positions of the full and dashed parts of the curve on a drawing indicate the hand of the locus.

Figure 16 shows the method for drawing a helix. First draw two views of the cylinder on which the helix is to be formed, making its length equal to the lead. Divide the lead and the circular view into the same number of equal parts (say twelve). When the point generating the curve has moved through one-twelfth of the circumference it has also moved through one-twelfth of the lead. Points on the curve can, therefore, be found by projecting point 1 on the circle to line 1 on the lead, and so on.

Helices occur in practice as screw threads, springs, worm gears, propeller blades and cylindrical cams, and examples of some of these are given on the following pages.

Figure 17 shows a circular section spring. The helical centre line of the spring is first drawn and at each plotted point on the helix a circle is drawn, having a diameter equal to the diameter of the wire. The outside profiles of the spring are drawn tangentially to these circles.

A square section spring is shown in Figure 18. Here two helices are present; one formed by the outside diameter and one by the inside diameter. Both helices, however, have the same lead. The construction should be obvious from the figure. Hidden detail is generally omitted on all drawings of helices or they become confusing.

Figure 19 shows a right-hand, single-start, square thread. Its construction is very similar to that of a square section spring except that the inner helices are modified by the presence of the solid core at the centre of the thread.

A helical chute is shown in Figure 20. The construction again is similar to that for a square section spring but is varied slightly by the cross-section of the chute.

Multi-start threads

If a quick-acting thread is required the lead must be large. This means that the thread depth will be large, and the cylinder on which the thread is cut will have a small core diameter and so will be weakened. To avoid this a multi-start thread may be used, such threads being two-start, three-start, etc. A common example is the thread on a fountain-pen cap. With multi-start threads the distance between corresponding

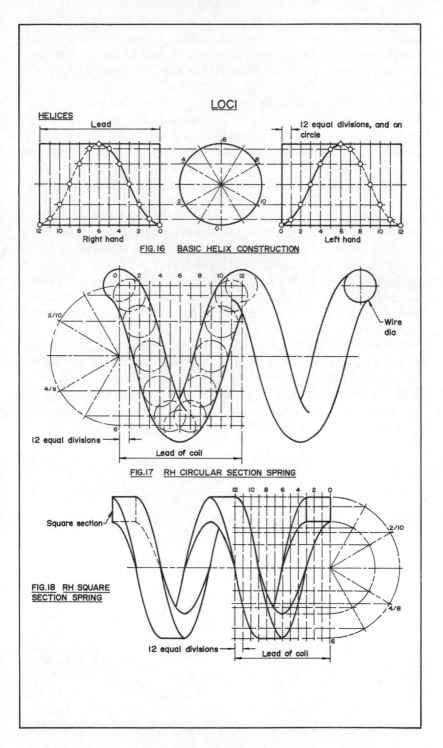

LOCI

HELICES

Lead

12 equal divisions, and on circle

6

4

8

2

10

0

12 10 8 6 4 2 0
Right hand

0 2 4 6 8 10 12
Left hand

FIG.16   BASIC HELIX CONSTRUCTION

0 2 4 6 8 10 12

Wire dia

2/10

4/8

6

12 equal divisions

Lead of coil

FIG.17   RH CIRCULAR SECTION SPRING

Square section

12 10 8 6 4 2 0

2/10

4/8

6

FIG.18 RH SQUARE SECTION SPRING

12 equal divisions

Lead of coil

33

points on adjacent threads is called the pitch, the lead being still the axial movement during one revolution. Thus, with a double-start thread the pitch is half the lead; with a triple-start thread it is a third of the lead, and so on. These points are illustrated in Figures 21 and 22.

The locus of a point on a mechanism

In the design of mechanisms it is frequently necessary to determine the locus of a point on the mechanism so that the forces present may be found and clearances checked. The locus may be plotted by drawing the mechanism in several positions and marking the position of the tracing point on each. A curve through these points will be the required locus.

Examples are given on page 37. Example 1 shows a crank and connecting rod mechanism. The end B of the connecting rod is constrained to move along the line CO, while the crank AO rotates about O. To plot the locus of P, divide the circle through A into a number of equal parts, twelve for convenience. From each point on the circle plot a corresponding position for B on CO, and draw AB in twelve positions. Plot P on each position of AB and, using a french curve, complete the locus as shown.

In Example 2 a rod AP always passes through a fixed point O, and A slides along a line VW. To plot the locus of P divide VW into a number of equal parts and draw AP at each of these divisions. As there is a large gap between P_3 and P_4, an extra point should be found between them to ensure that the locus is accurate. Join the plotted positions of P, using a french curve, to complete the locus as shown.

LOCI

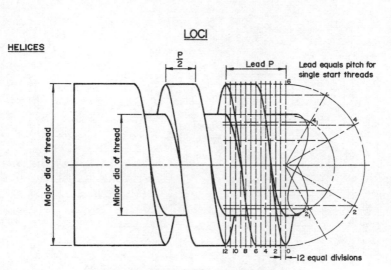

Lead equals pitch for single start threads

FIG.19   RH SINGLE START SQUARE THREAD

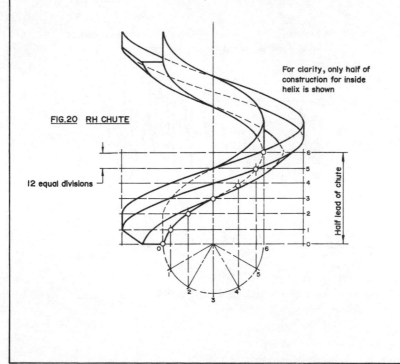

For clarity, only half of construction for inside helix is shown

FIG.20   RH CHUTE

12 equal divisions

Half lead of chute

35

LOCI

HELICES

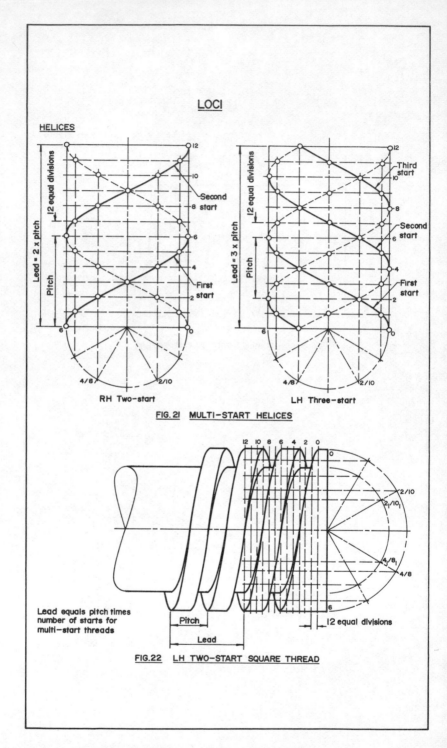

FIG. 21 MULTI-START HELICES

RH Two-start

LH Three-start

Lead = 2 x pitch

Lead = 3 x pitch

Lead equals pitch times
number of starts for
multi–start threads

Pitch

Lead

12 equal divisions

FIG.22 LH TWO-START SQUARE THREAD

36

LOCI

MECHANISMS

Example I. OA revolves about O, B slides along CO. Plot the locus of P.

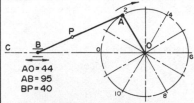

AO = 44
AB = 95
BP = 40

Stage I. Draw mechanism. Divide circle through A into 12 equal parts.

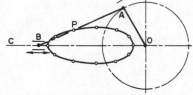

Stage 2. Draw AB for each of the 12 positions of A.

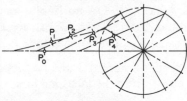

Stage 3. Plot P on each position of AB.

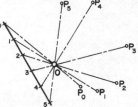

Stage 4. Complete locus using french curve.

Example 2. A slides between V and W. AP always passes through O. Plot the locus of P.

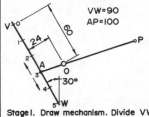

VW = 90
AP = 100

30°

Stage I. Draw mechanism. Divide VW into a number of equal parts.

Stage 2. Draw AP at each of the divisions on VW.

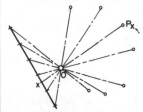

Stage 3. Plot extra points P_x to fill any gaps in curve.

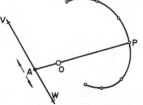

Stage 4. Complete locus using french curve.

37

1 Draw an ellipse with a major axis of 150 mm and a minor axis of 100 mm by the focal point method.

2 Draw an ellipse having axes of 120 mm and 80 mm by the auxiliary circles method.

3 An approximate ellipse is to have axes of 166 and 108 mm. Draw the figure.

4 Construct a parabola using the locus method, which has its vertex 20 mm from the directrix.

5 An ellipse, parabola and hyperbola have a common focus 40 mm from the directrix. Draw all three curves if the eccentricity of the ellipse is 2/3 and of the hyperbola 3/2.

6 Draw a parabola with its axis vertical, in a rectangle 128 mm high by 114 mm wide.

7 Two lines OA and OB are at right angles with OA horizontal and OB vertical. OA is 150 mm long and OB 130 mm. Construct a parabola to pass through A and B using the envelope method.

8 A hyperbola has its vertex 12 mm from the directrix and its focus 20 mm from the vertex. Draw the curve using the locus method.

9 Draw a rectangular hyperbola having a point on the curve 18 mm from the vertical asymptote and 76 mm from the horizontal asymptote.

10 Using the construction shown in Figure 9, page 26, draw both branches of a hyperbola with a transverse axis of 60 mm and a point on one branch 60 mm from the axis of symmetry and 72 mm from the vertex.

11 Draw a cycloid for a rolling circle 76 mm diameter, making the initial position of the tracing point at the top of the vertical centre line of the circle.

12 Draw a superior trochoid for a point 12 mm outside a rolling circle of 64 mm diameter. Let the circle make 1.5 revolutions with the initial position of the tracing point at the bottom of its vertical centre line.

13 A point P is 16 mm inside a 76 mm diameter circle. If the circle rolls for one revolution along a fixed horizontal line, draw the locus of P. P is to start in its highest position.

14 Draw as shown in Figure 13, page 30, an epicycloid and hypo-cycloid for 75 mm diameter rolling circles and a 225 mm diameter base circle.

15 Draw a regular pentagon with sides 30 mm long and construct its involute.

16 Construct three-quarters of a turn of an involute to a 50 mm diameter base circle.

17 Draw one complete turn for right-hand and left-hand helices 44 mm diameter and 72 mm lead.

18 Draw two complete turns of a right-hand helical spring, 76 mm mean diameter, 48 mm pitch, made from 16 mm diameter wire.

19 Draw two complete turns of a left-hand helical spring, 120 mm outside diameter and 36 mm pitch, made from 15 mm square section wire.

20 Draw a 120 mm length of a right-hand, single-start square thread, 100 mm outside diameter and 36 mm pitch.

21 Draw a 150 mm length of a left-hand, two-start square thread, 120 mm outside diameter and 84 mm lead.

22 In the given mechanism on page 41, the crank OA revolves anti-clockwise about O. The end B of the rod AB is constrained to move always along PQ. Plot the locus of R for one revolution of OA if OA is 30 cm, AB is 105 cm and AR is 70 cm. Scale 1 mm to 1 cm.

23 The crank OA revolves anticlockwise about O, and B moves to and fro along the horizontal line through O. Plot the locus of P if OA is 30 mm, AB is 95 mm and AP is 20 mm.

24 The rod AB moves so that A is always on OY and B is always on OX. Plot the locus of P for the maximum movements of A and B if AB is 130 mm and AP is 58 mm.

25 The figure shows diagrammatically a pair of folding doors. Plot the locus of P for the full movement of A from D to C. AB and BC are each 150 cm and AP is 60 cm. Scale 1 mm to 1 cm.

26 In the mechanism shown in the figure, OA revolves anti-clockwise about O while AB slides through the pivoted block C. Draw the locus of B for one revolution of OA. OA is 40 mm and AB is 145 mm.

27 In the given mechanism the cranks AO and BQ revolve in opposite directions at the same speed, and are joined by the rods AC and BCP. Plot the locus of P for one revolution of the cranks, if AO and BQ are 25 mm, AC is 125 mm and CP is 20 mm.

28 The crank OA of the mechanism shown rotates clockwise about O. The end B of the link AB moves along the line PQ and FD swings about F. Obtain the locus of E for one revolution of OA. OA is 40 mm, AB is 150 mm, BC is 65 mm, CD is 130 mm, DE and DF are 75 mm.

29 In the figure the crank AB rotates clockwise about A and the crank CD rotates anticlockwise about C. The cranks are joined by the link BDF. Plot to a scale of 1 mm to 1 cm the loci of points E and F for one revolution of AB. AB and DC are 45 cm, BD is 120 cm and BE and DF are 30 cm.

30 The crank OA in the figure rotates clockwise at constant speed and during one revolution E moves from C to D and back to C at constant speed. If the initial position of the mechanism is as shown in the figure, draw the locus of B during one revolution of OA. OA is 25 mm and AB and BC are 75 mm.

31 The rod OA rotates anticlockwise at constant speed about O, through 180°. During this movement the point P moves from P to P_1 and back to P at constant speed. Draw the locus of P during the complete movement of OA.

32 The crank in the figure rotates clockwise about O. The link AB is attached to a rod CB which swings about C. Construct the locus of P if OA is 45 cm, AB is 180 cm, CB is 75 cm and BP is 90 cm. Scale 1 mm to 1 cm.

33 A modified form of Watt's straight line motion is shown in Figure 33. The rods AB and CD are 115 mm long and swing about A and D respectively. The link BC is 76 mm long with P its mid point. Plot the complete locus of P.

LOCI PROBLEMS

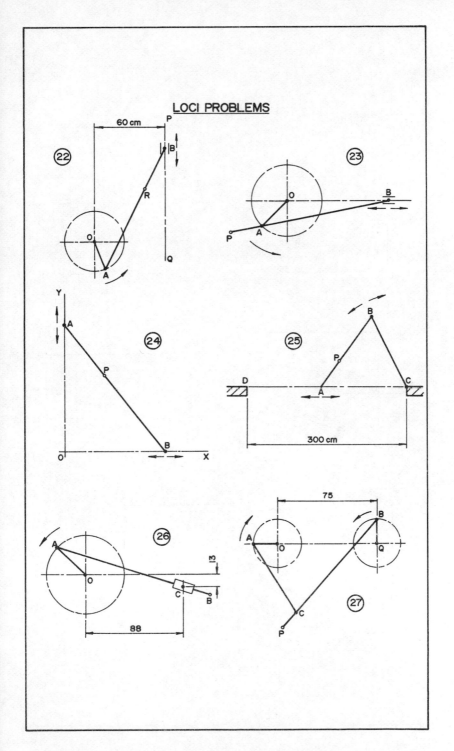

41

LOCI PROBLEMS

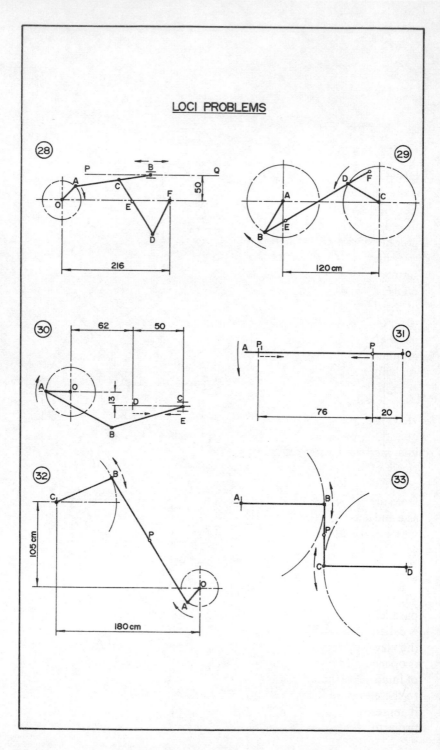

5

ORTHOGRAPHIC PROJECTION

FIGURE 1 on page 45 shows a point A in space and a plane. If a line is drawn from the point to the plane meeting it at A', then A' is a projection of A on the plane. The line AA' is a projector. If the projector is normal to the plane, A' is an orthographic projection of A – 'orthographic' meaning 'to draw at right angles'.

A line may be projected orthographically by projecting its ends as shown in Figure 2. The projectors will be parallel and normal to the plane. It should be noted that the projection will show the true length of the line only if the line is parallel to the plane.

Solids consist of points linked by lines. Therefore, using systems of parallel projectors from their boundaries, they may be projected orthographically on to any number of planes. This is illustrated in Figure 3. As before, the projectors are normal to the planes.

In practice, orthographic projection uses two main planes, called the principal planes of projection. One is horizontal, the other vertical, and views on them are principal views. They intersect producing four quadrants or angles as shown in Figure 4. The object to be drawn is imagined to be placed in one of these quadrants, and orthographic views of it are projected on to the planes. The object may have any orientation to the planes, but normally it is positioned so that its main faces are parallel to them. This ensures that views of the faces are true size and shape. In practice only the First and Third Angles are used since views in the Second and Fourth Quadrants may overlap.

First Angle orthographic projection
Figure 5 shows an object positioned in space in the First Quadrant. Views of the object have been drawn on the planes using systems of parallel projectors normal to the planes. The view on the vertical plane is called the elevation, that on the horizontal plane the plan. To obtain the views as they would appear on a sheet of paper, the horizontal plane is opened out, or rabatted, about the intersection of the planes. The line of intersection is called the XY line, ground line or folding line. Relative to the elevation it represents the horizontal plane. Relative to the plan it represents the vertical plane. It will be seen from Figure 5 that the projectors cross the ground line at right angles and are parallel. Since

the ground line represents the planes, the views in Figure 5 satisfy the conditions for orthographic projection given above.

End views in First Angle projection

It is found in practice that an elevation and plan of an object are not always sufficient to describe it completely. If this is the case a third view, called an end view, end elevation or side elevation is drawn on an auxiliary vertical plane. This plane, as shown in Figure 6, is at right angles to both the horizontal and vertical planes. It may be placed in either of the two positions shown, depending on which face of the object is the more important. If neither face is more important than the other, the end view to be drawn is that which shows the minimum of hidden lines. In some cases it may be necessary to draw both end views.

When the auxiliary vertical plane is opened out with the horizontal plane, the views appear in the positions given in Figure 6. It will be seen that heights in the two elevations are equal and depths in the plan view equal corresponding widths in the end view. These points should be borne in mind when orthographic views are being drawn, since they make it possible to project the third view from any two given views.

Some objects have inclined or oblique faces. An inclined face makes an angle with one principal plane and is perpendicular to the other; an oblique face is at an angle to both principal planes. To describe completely objects of these types the three views already discussed are usually insufficient, and it is necessary to project additional views on auxiliary planes. This aspect of orthographic projection is dealt with in Book 2.

Projection symbols

Since, as BS 308:1972 states, 'two systems of projection, First Angle and Third Angle are approved internationally', it is necessary to indicate on the drawing which system has been used. This is done by a symbol consisting of an elevation and end view of a frustum of a cone. The First Angle symbol is shown in Figure 6 and the Third Angle symbol in Figure 8. If the symbols are not used then BS 308:1972 recommends that 'the direction in which the views are taken should be clearly indicated'. Failure to indicate the projection system may result in costly errors being made by a person, used to one system, misreading a drawing made in the other.

Third Angle orthographic projection

An object positioned in the Third Quadrant between the principal planes is shown in Figure 7. Since the planes now come between the observer and the object they are imagined to be transparent, and the object is viewed through them. An elevation and plan have been projected on to the vertical and horizontal planes respectively, using parallel projectors normal to the planes as for First Angle projection.

ORTHOGRAPHIC PROJECTION

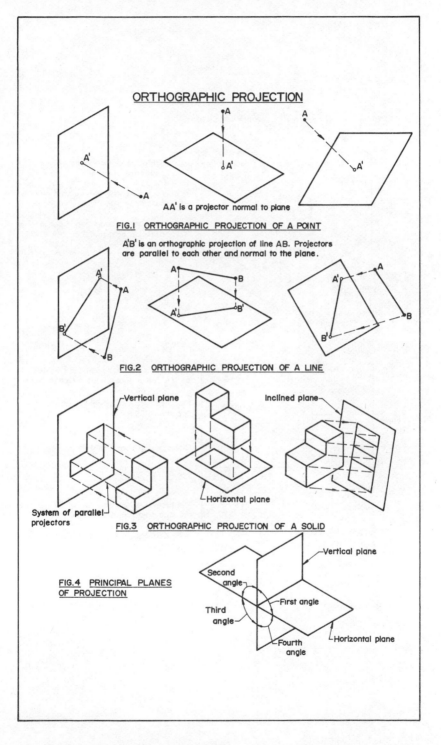

AA' is a projector normal to plane

FIG.1   ORTHOGRAPHIC PROJECTION OF A POINT

A'B' is an orthographic projection of line AB. Projectors are parallel to each other and normal to the plane.

FIG.2   ORTHOGRAPHIC PROJECTION OF A LINE

Vertical plane

Inclined plane

Horizontal plane

System of parallel projectors

FIG.3   ORTHOGRAPHIC PROJECTION OF A SOLID

FIG.4   PRINCIPAL PLANES OF PROJECTION

Vertical plane

Second angle

First angle

Third angle

Fourth angle

Horizontal plane

45

ORTHOGRAPHIC PROJECTION

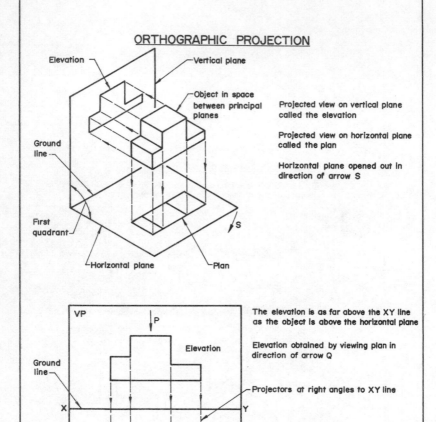

Elevation

Vertical plane

Object in space
between principal
planes

Ground
line

First
quadrant

Horizontal plane

Plan

S

Projected view on vertical plane
called the elevation

Projected view on horizontal plane
called the plan

Horizontal plane opened out in
direction of arrow S

VP

P

Elevation

Ground
line

X

Y

Plan

HP

Q

The elevation is as far above the XY line
as the object is above the horizontal plane

Elevation obtained by viewing plan in
direction of arrow Q

Projectors at right angles to XY line

The plan is as far below the XY line as
the object is from the vertical plane

Plan view obtained by viewing elevation
in direction of arrow P

FIG.5   PRINCIPAL VIEWS IN FIRST ANGLE ORTHOGRAPHIC PROJECTION

46

ORTHOGRAPHIC PROJECTION

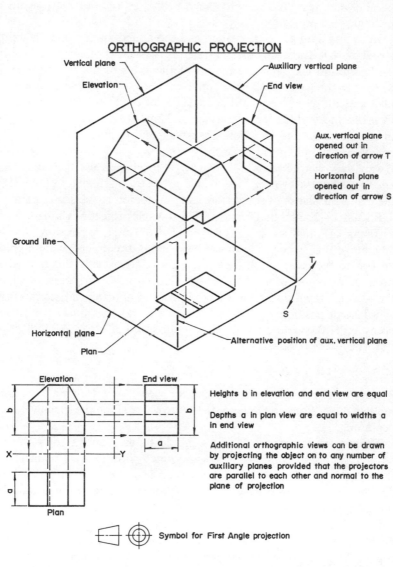

Vertical plane

Auxiliary vertical plane

Elevation

End view

Aux. vertical plane opened out in direction of arrow T

Horizontal plane opened out in direction of arrow S

Ground line

T

S

Horizontal plane

Alternative position of aux. vertical plane

Plan

Elevation

End view

Heights b in elevation and end view are equal

Depths a in plan view are equal to widths a in end view

Additional orthographic views can be drawn by projecting the object on to any number of auxiliary planes provided that the projectors are parallel to each other and normal to the plane of projection

X — Y

Plan

Symbol for First Angle projection

FIG.6 PROJECTION OF END VIEWS IN FIRST ANGLE ORTHOGRAPHIC PROJECTION

47

When the planes are rabatted the views appear as in Figure 7, with the plan now above the elevation. Figure 7 shows the projectors to be parallel and at right angles to the XY line, so the views are again in orthographic projection.

In Figure 8 an auxiliary plane has been added and an end view projected on to it. Rabatting the planes gives the views positioned as shown. The plan and end view are placed at the sides of the front elevation nearest to the faces which they represent. In First Angle projection the plan and end view are placed at the sides of the front elevation remote from the faces which they describe. In Third Angle as in First, heights in the elevation and end view are equal, and depths in the plan view equal corresponding widths in the end view.

Auxiliary views drawn in Third Angle projection use the same principles as those drawn in First Angle projection. Auxiliary planes, like the principal planes of projection, are considered to be transparent.

BS 308:1972 states that First and Third Angle projections 'are regarded as being of equal status'. However, Third Angle projection has the advantage that, when long objects are being drawn, end views appear nearest to the faces which they represent. Occasionally this is used to produce a combination of First and Third Angle projections in which the plan is drawn in the First Angle position and the end views in the Third Angle position. Such mixed-projection drawings must always carry notes under the views such as 'view in direction of arrow A', indicating how they were obtained.

General
Always leave enough space between the views on a drawing to accommodate dimensions and notes without crowding. Plan the spacing before beginning.

The minimum number of views should be used consistent with describing the object completely. A view which shows only a diameter or thickness is unnecessary if this information can be given as a note or a dimension on another view. A view without a note or dimensions is probably unnecessary.

Hidden detail should only be used where it is essential for a complete description of the object, but it should not be used for dimensioning. Figure 9 shows the correct representation of hidden detail in various situations.

Build up all the views together. Completing the views separately wastes time since measurements can often be made on two or more views simultaneously, or projected from one to another as soon as they are made.

Three worked examples in each system of projection follow on pages 52 to 55 and further problems will be found on page 56.

ORTHOGRAPHIC PROJECTION

Plan

Horizontal plane

Ground line

S

Third quadrant

Horizontal plane opened out in direction of arrow S

Vertical plane

In Third Angle projection planes are transparent and object is viewed through them

Elevation

Object in space between principal planes

HP

Plan

The plan is as far above the XY line as the object is from the vertical plane

Plan view obtained by viewing elevation in direction of arrow P

Q

Projectors at right angles to XY line

X   Y

P

The elevation is as far below the XY line as the object is below the horizontal plane

Elevation

Elevation obtained by viewing plan in direction of arrow Q

VP

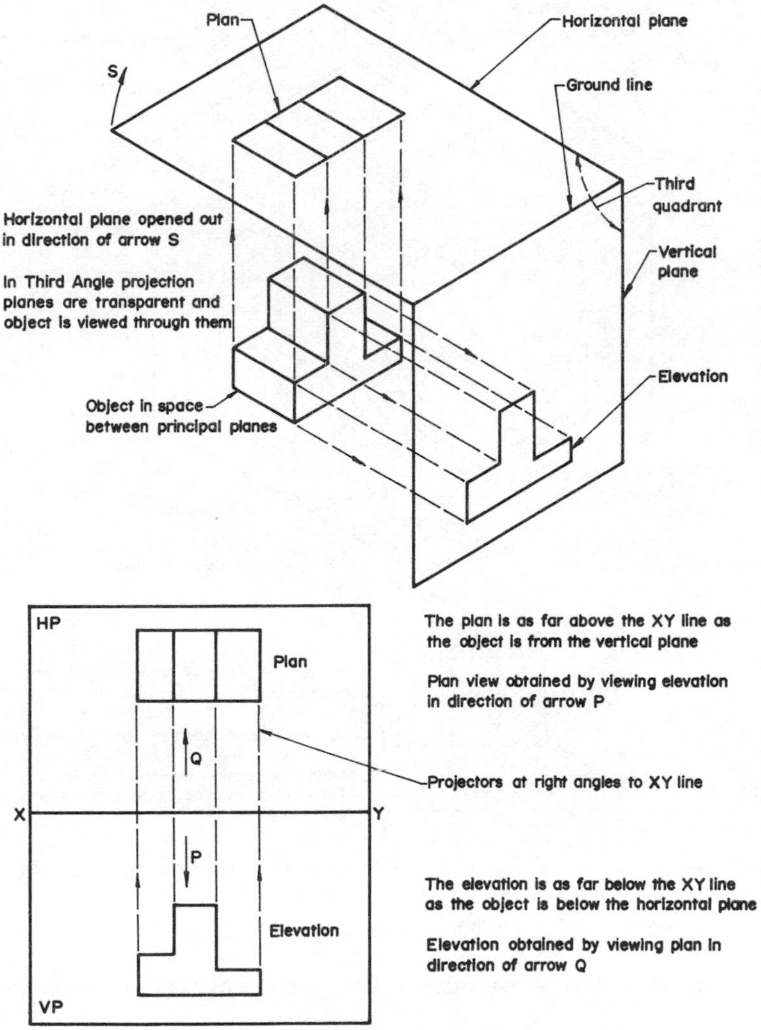

FIG.7   PRINCIPAL VIEWS IN THIRD ANGLE ORTHOGRAPHIC PROJECTION

ORTHOGRAPHIC PROJECTION

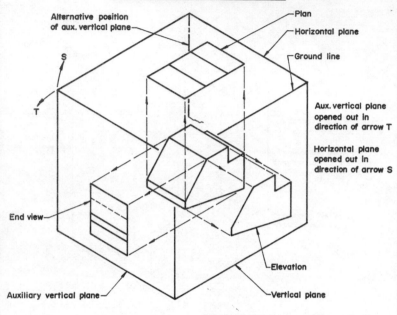

Alternative position of aux. vertical plane

Plan

Horizontal plane

Ground line

S

T

Aux. vertical plane opened out in direction of arrow T

Horizontal plane opened out in direction of arrow S

End view

Elevation

Auxiliary vertical plane

Vertical plane

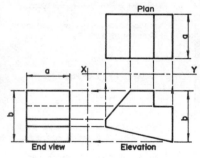

Plan

a

X | Y

a

b

b

End view

Elevation

Depths a in plan view are equal to widths a in end view

Heights b in elevation and end view are equal

Additional auxiliary views can be drawn in Third Angle projection in the same way as in First Angle

 Symbol for Third Angle projection

FIG.8   PROJECTION OF END VIEWS IN THIRD ANGLE ORTHOGRAPHIC PROJECTION

50

ORTHOGRAPHIC PROJECTION

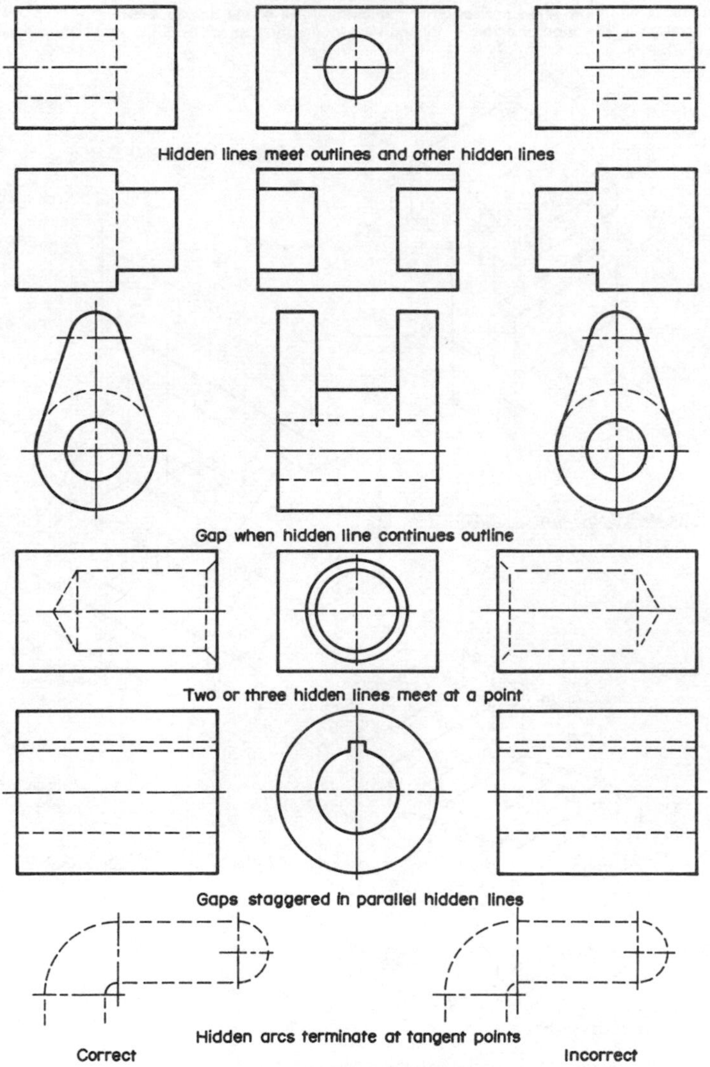

Hidden lines meet outlines and other hidden lines

Gap when hidden line continues outline

Two or three hidden lines meet at a point

Gaps staggered in parallel hidden lines

Hidden arcs terminate at tangent points

Correct   Incorrect

FIG.9   HIDDEN LINE TECHNIQUE

ORTHOGRAPHIC PROJECTION EXAMPLES

Draw full size in First Angle projection the following views of the details shown.
(a) Elevation in direction of arrow T   (b) End view in direction of arrow S   (c) Plan view projected from view (a)

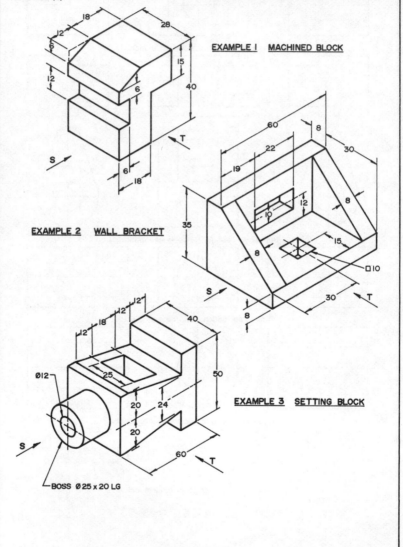

EXAMPLE I   MACHINED BLOCK

EXAMPLE 2   WALL BRACKET

EXAMPLE 3   SETTING BLOCK

BOSS Ø25 x 20 LG

ORTHOGRAPHIC PROJECTION EXAMPLES

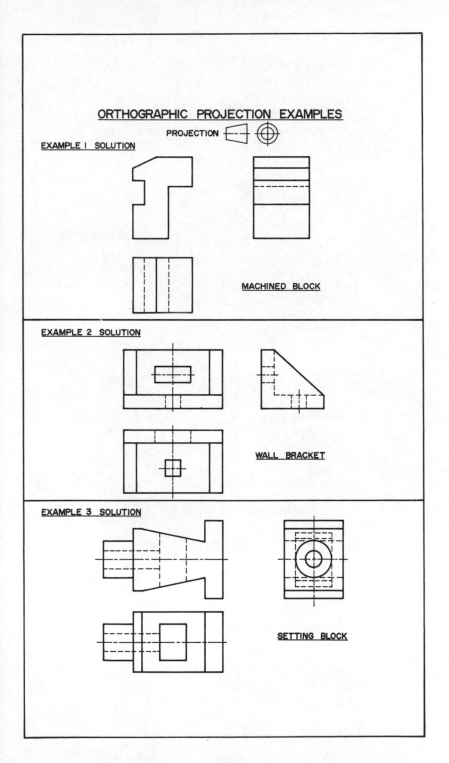

PROJECTION

EXAMPLE 1 SOLUTION

MACHINED BLOCK

EXAMPLE 2 SOLUTION

WALL BRACKET

EXAMPLE 3 SOLUTION

SETTING BLOCK

ORTHOGRAPHIC PROJECTION EXAMPLES

Draw full size in Third Angle projection the following views of the details shown.
(a) Elevation in direction of arrow T (b) End view in direction of arrow S (c) Plan view projected from view (a)

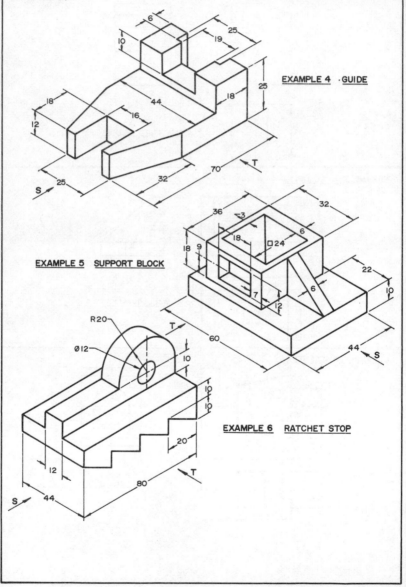

EXAMPLE 4 ·GUIDE

EXAMPLE 5   SUPPORT BLOCK

EXAMPLE 6   RATCHET STOP

54

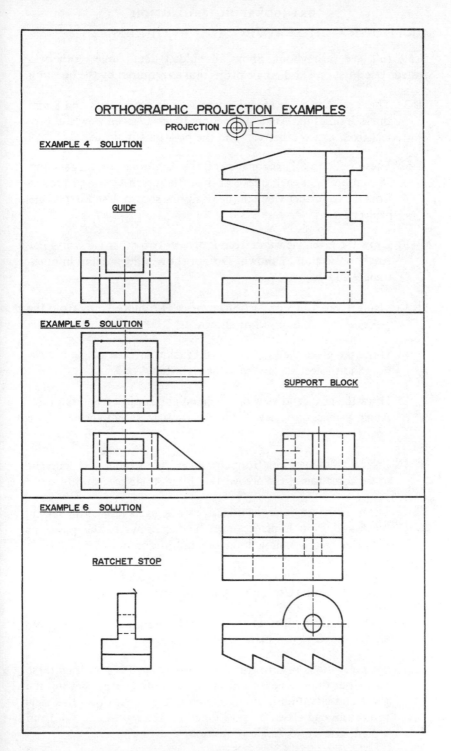

ORTHOGRAPHIC PROJECTION EXAMPLES

PROJECTION

EXAMPLE 4 SOLUTION

GUIDE

EXAMPLE 5 SOLUTION

SUPPORT BLOCK

EXAMPLE 6 SOLUTION

RATCHET STOP

ORTHOGRAPHIC PROJECTION PROBLEMS

Scale full size throughout. Show all hidden detail unless otherwise stated. Use First or Third Angle projection as required by the question.

1–3 These refer to a hexagonal prism 24 mm across flats and 8 mm thick. Using First Angle projection draw the given view and project from it two other views in the positions indicated.

4–6 These refer to a hexagonal pyramid 30 mm across corners and 28 mm high. In each question draw the given view and project from it two further views in the positions shown. Use First Angle projection.

7–10 Draw the given view and from it project two other views in First Angle projection as shown. Do not show hidden detail in questions 9 and 10.

11–13 Using First Angle projection, project two more views from the given view in the positions shown.

14 Draw the given views of the wall bracket in First Angle projection and project a plan from the right-hand one.

15 Draw the given end view of the detail and project from it in First Angle projection a new front elevation in the direction of the arrow, and a plan.

16 Project from the given front elevation a plan and a new end view in the direction of the arrow. Use First Angle projection.

17–20 Using Third Angle projection, draw the given view of the setting piece and from it project two further views in the positions indicated. Do not show hidden detail in questions 19 and 20.

21–23 Draw the given view in each case and project from it two other views as shown. Use Third Angle projection.

24 Copy the given views of the camshaft bearing and project from the left-hand view a plan. Use Third Angle projection.

25 Two elevations of a bearing base are given. Do not draw these views but draw a new front elevation obtained by viewing the given end elevation in the direction of arrow A. From this view project an end elevation positioned on the right-hand side, and a plan. Use Third Angle projection.

26 Draw a new front elevation of the clamp, obtaining it by viewing the given end view in the direction of arrow A. From this view project in Third Angle projection an end view positioned on the left-hand side, and a plan.

Further simple problems in orthographic projection will be found at the beginning of the Machine Drawings in Chapter 17.

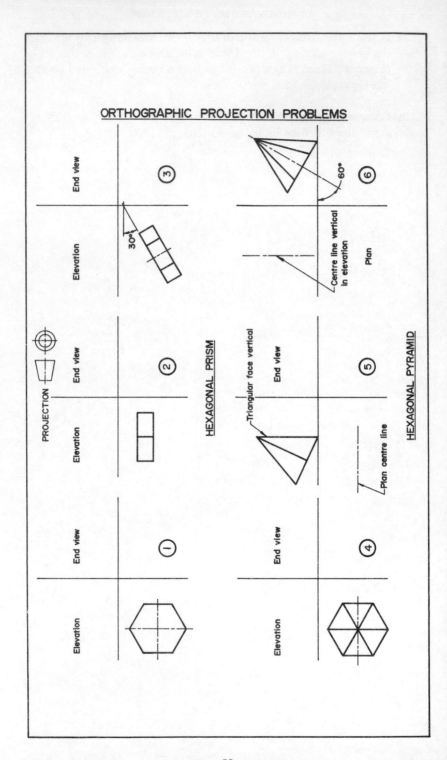

ORTHOGRAPHIC PROJECTION PROBLEMS

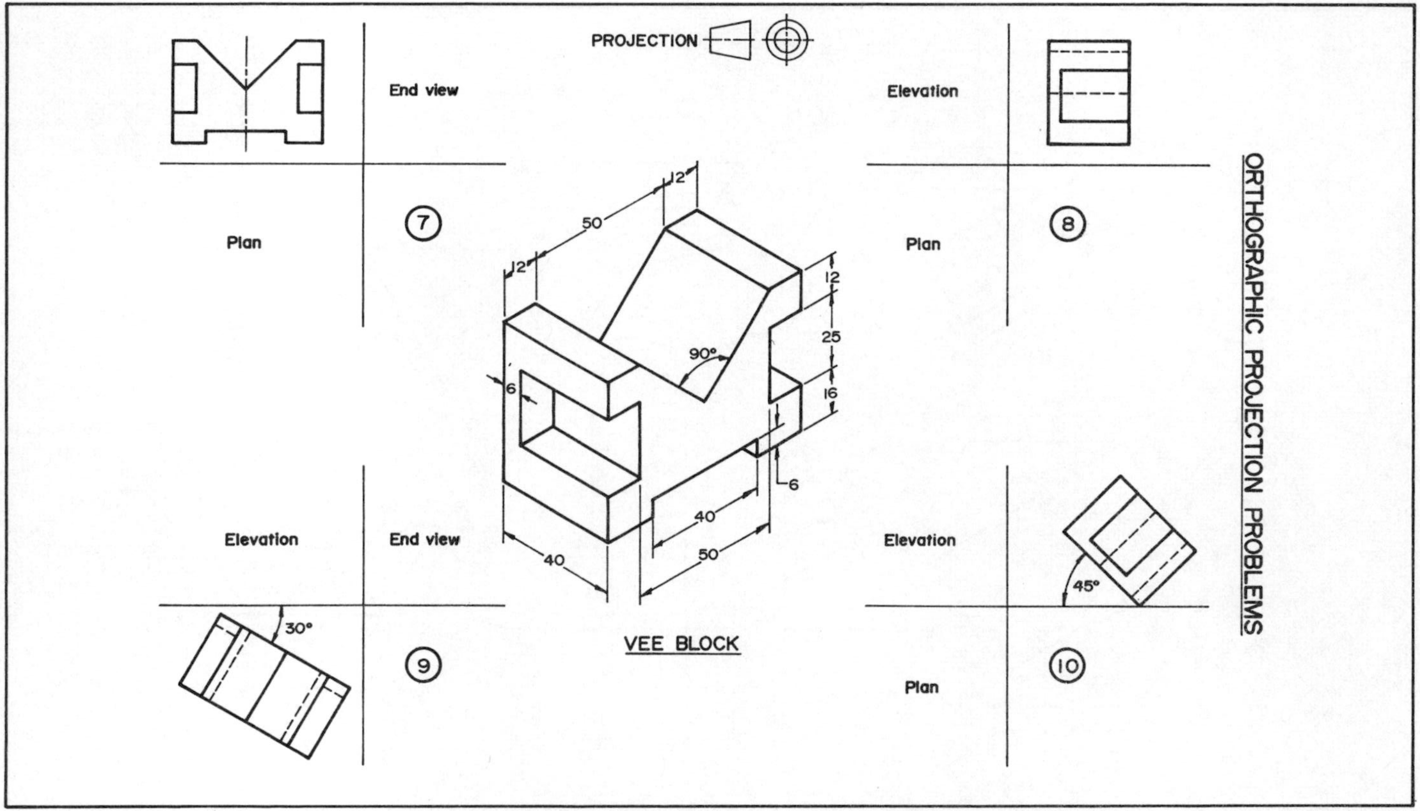

PROJECTION

VEE BLOCK

(7)

(8)

(9)

(10)

Elevation

Plan

Elevation

Plan

End view

Plan

End view

Elevation

45°

30°

90°

12

25

16

6

40

50

12

50

12

6

40

ORTHOGRAPHIC PROJECTION PROBLEMS

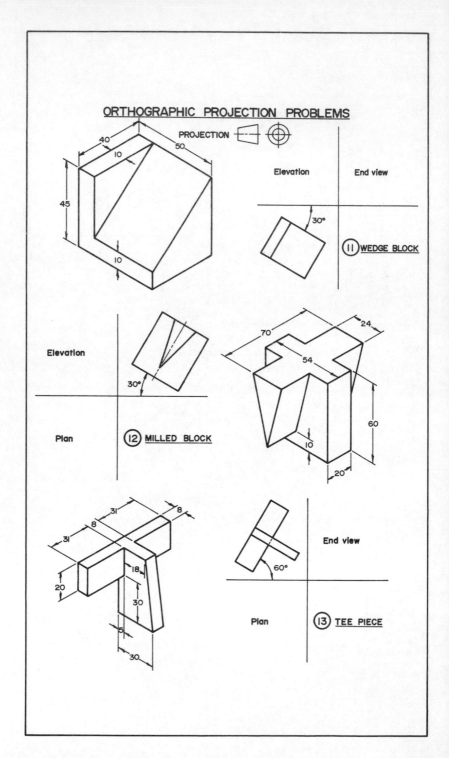

PROJECTION

Elevation   End view

(11) WEDGE BLOCK

Elevation

Plan   (12) MILLED BLOCK

End view

Plan   (13) TEE PIECE

ORTHOGRAPHIC PROJECTION PROBLEMS

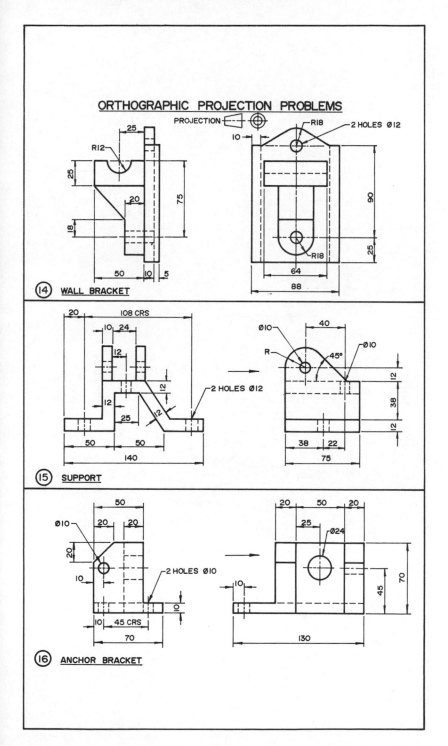

PROJECTION

(14) WALL BRACKET

25

R12

25

20

18

75

50   10   5

R18   2 HOLES Ø12

10

90

25

R18

64

88

(15) SUPPORT

20   108 CRS

10   24

12

12

12

2 HOLES Ø12

12

25   12

50   50

140

Ø10   40

R   45°   Ø10

12

38

12

38   22

75

(16) ANCHOR BRACKET

50

Ø10   20   20

20

10   2 HOLES Ø10

10

10   45 CRS

70

20   50   20

25   Ø24

10

70

45

130

61

ORTHOGRAPHIC PROJECTION PROBLEMS

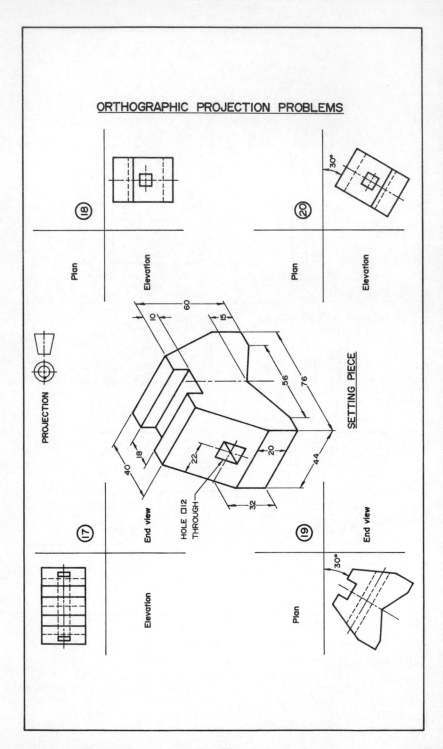

SETTING PIECE

HOLE □12 THROUGH

PROJECTION

⑰ End view / Elevation

⑱ Plan / Elevation

⑲ End view / Plan

⑳ Plan / Elevation

ORTHOGRAPHIC PROJECTION PROBLEMS

PROJECTION

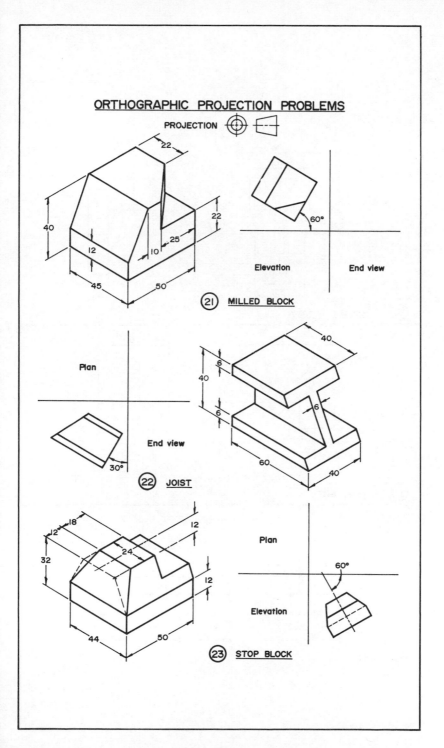

Elevation   End view

21 MILLED BLOCK

Plan

End view

22 JOIST

Plan

Elevation

23 STOP BLOCK

63

ORTHOGRAPHIC PROJECTION PROBLEMS

PROJECTION

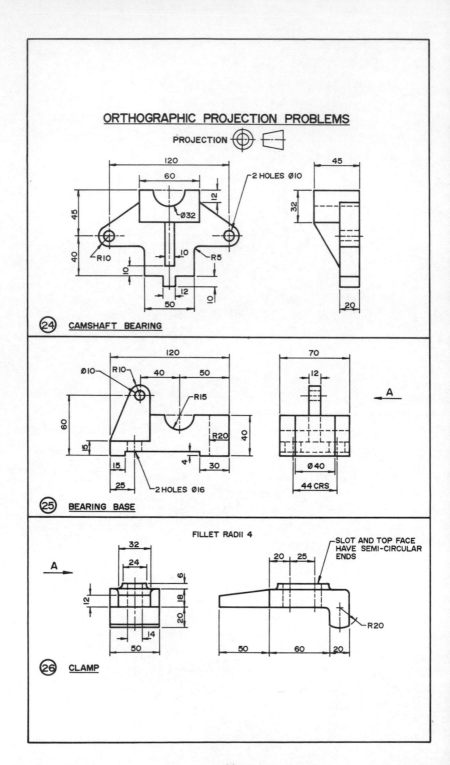

(24) CAMSHAFT BEARING

(25) BEARING BASE

(26) CLAMP

FILLET RADII 4

SLOT AND TOP FACE HAVE SEMI-CIRCULAR ENDS

6

DIMENSIONING

THE study of dimensioning falls into two parts. First, what may be called the technique of dimensioning, that is, the correct drawing, lettering and positioning of the dimensions on a drawing. This may be quickly grasped by learning the various rules and conventions which govern it, and by studying finished drawings in this book and elsewhere. Secondly, dimensioning involves the selection of dimensions to ensure the correct functioning of the part, and to enable the workman to make it without having to calculate any sizes. This side of dimensioning is only understood after considerable experience both in the making of drawings, and in the workshop processes by which the article is made. There are, however, some basic principles for the selection of dimensions which are set out below.

Dimensioning technique
Dimensions should be placed outside the outline of the view wherever possible. This is achieved by drawing projection or extension lines from points or lines on the view and placing a dimension line between them. Dimension and projection lines are thin, continuous lines as noted in Chapter 1. There should preferably be a small gap between the outline and the start of the projection line, and projection lines should continue a short distance beyond the dimension line. The dimension line has arrowheads about 3 mm long at each end, and these must just touch the projection or other limiting line. The dimension line which is nearest the outline should be about 10 mm from it if possible, and succeeding dimensions should be well spaced for clarity. Dimensions must never be cramped. To avoid dimension and projection lines crossing, the smallest dimensions should be placed nearest the outline. These points are shown in Figure 1 on page 66.

Centre lines or their extensions, and outlines or their extensions, must not be used as dimension lines. They may, however, be used as projection lines as shown in Figure 2. Dimension lines should be placed on the view which shows the features to which they refer most clearly.

Dimension figures are placed normal to the dimension line and near its centre. They must not be crossed or separated by another line of the drawing. The figures must be positioned so that they can be read from

DIMENSIONING

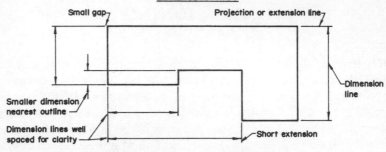

Small gap

Projection or extension line

Dimension line

Smaller dimension nearest outline

Dimension lines well spaced for clarity

Short extension

FIG.I   DIMENSION AND PROJECTION LINES

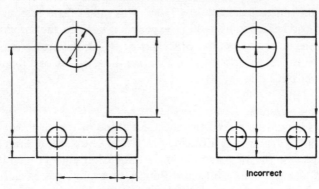

Incorrect

FIG.2   PLACEMENT OF DIMENSION LINES

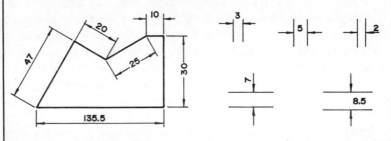

20

10

47

25

30

135.5

3   5   2

7

8.5

FIG.3 ALIGNMENT OF DIMENSIONS   **FIG.4 DIMENSIONING OF SMALL FEATURES**

66

the bottom or right-hand side of the drawing, either above the dimension line or in a gap in it. Figure 3 illustrates these points. Figure 4 shows how small features or narrow spaces are dimensioned. The dimension is placed centrally or above the extension of one of the arrowheads.

The decimal marker should be bold and placed on the base line of the figures. Dimensions less than unity should be preceded by a zero.

The dimension line for an angle is a circular arc having its centre on the point of the angle. This is illustrated in Figure 5. The dimension figures, like those for linear dimensions, are placed so that they can be read from the bottom or right-hand side of the drawing. Angular dimensions are given in degrees, degrees and minutes, or degrees, minutes and seconds, depending on the accuracy required. If the angle is less than one degree it should be preceded by $0°$.

Complete circles must always be dimensioned by their diameters, using one of the methods shown in Figure 6. The dimension is preceded by the symbol ϕ, meaning diameter. Circles must be shown with two centre lines.

Figure 7 shows several ways in which diameters may be dimensioned. The dimension should be placed on the view which ensures the maximum clarity, as at (a). Here the diameter dimensions are placed on the longitudinal view rather than on the view having a number of concentric circles. At (b) the diameter dimensions are related to the features by leaders, thus avoiding a large number of projection lines appearing on the view. The methods at (c) and (d) are useful where space is restricted.

Radii are dimensioned using a dimension line which passes through, or is in line with, the arc centre. The dimension line carries one arrowhead only and this touches the arc. The abbreviation R precedes the dimension. Figure 8 illustrates these points.

Methods of dimensioning chamfers are shown in Figure 9. To avoid any misinterpretation of the dimensions of $45°$ chamfers one of these methods should be used, and not a leader and note.

Notes are frequently used on drawings together with a leader which shows where the note applies. Leaders are thin, continuous lines which may terminate in an arrowhead or dot, as shown in Figure 10. Leaders with arrowheads must touch a line; those with dots should have the dot within the outline. Leaders must not touch an outline at an acute angle, neither should they be parallel to adjacent dimension or projection lines or confusion may occur. Long leaders should be avoided either by placing the note or dimension close to the view, or by repeating it.

Recommended ways of dimensioning the common machining operations of countersinking, counterboring and spotfacing are shown on page 72. Notes such as 'Countersink to suit M8 Csk Hd screw' are imprecise and should be avoided. It should be noted that since spotfacing is an operation which only cleans up a surface, a depth dimension is unnecessary.

The dimensioning of screw threads and keyways is covered in Chapters 9 and 12 respectively.

The foregoing notes are based on BS 308: Part 2: 1972, Dimensioning and tolerancing of size. They are intended to cover only those points which occur frequently, and for a more detailed treatment students should consult the Standard.

The selection of dimensions—size dimensions

Engineering details are made up of simple geometrical shapes such as cylinders, prisms, cones, spheres and so on. These shapes may be 'positive' or 'negative'. For example, a circular pin is a positive cylinder, while a circular hole is a negative cylinder. When dimensioning a detail, first break it down into its simple shapes and dimension them to show their sizes. These are the size dimensions. Then position the simple shapes relative to each other by location dimensions.

Examples of the size dimensioning of the more common geometrical shapes are given in Figure 14. The cylinder and prism need to have only their length and cross-section stated. Note, however, that for a negative cylinder in the form of a blind drilled hole, the length is the depth of the circular portion, not the depth to the point, since this cannot be easily measured. The complete cone, a rarity in engineering, requires the vertical height and base diameter. For a frustum of a cone one of the other methods is used. Spherical diameters or radii should be preceded by 'Sphere'.

Location dimensions

When location dimensions are selected the following points should be borne in mind. They are illustrated in Figure 15. Cylinders, cones and other symmetrical shapes are located by their centre lines and not by their surfaces. Non-symmetrical shapes are located by their corners or faces. The location dimensions should appear on the view which shows the 'typical shape' of the feature. For example, holes should be located on the view which shows them as circles. Holes equally spaced on a circle (a 'pitch circle') should be so described in the note giving their number and size, or with a separate dimension for the pitch circle diameter. If they are not equally spaced they should be located either by angles or ordinates, depending on the accuracy required.

Castings and forgings are generally only partially machined. For such parts location dimensions must be taken from the machined surfaces or from centre lines. Unmachined surfaces must not be used. Where it is necessary to indicate that a surface is to be machined, without defining either the surface texture, grade or process to be used, a symbol of the form shown in Figure 15 should be used. The symbol should be applied normal to the line representing the surface, or to a leader of

DIMENSIONING

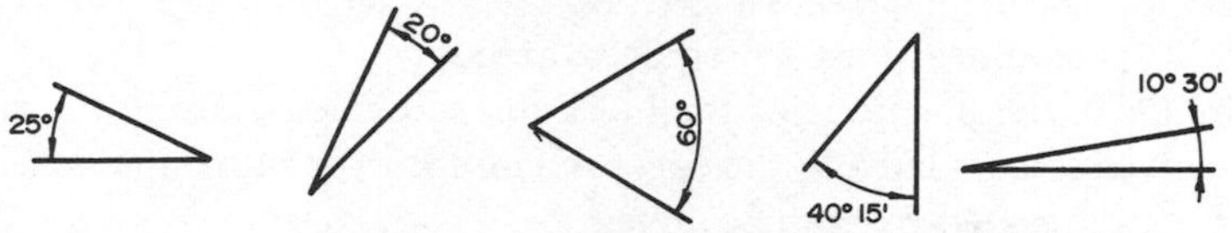

FIG.5   DIMENSIONING ANGLES

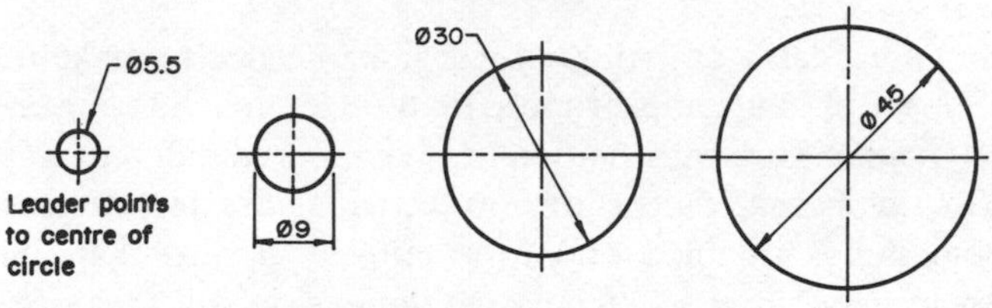

Leader points
to centre of
circle

FIG.6   DIMENSIONING CIRCLES

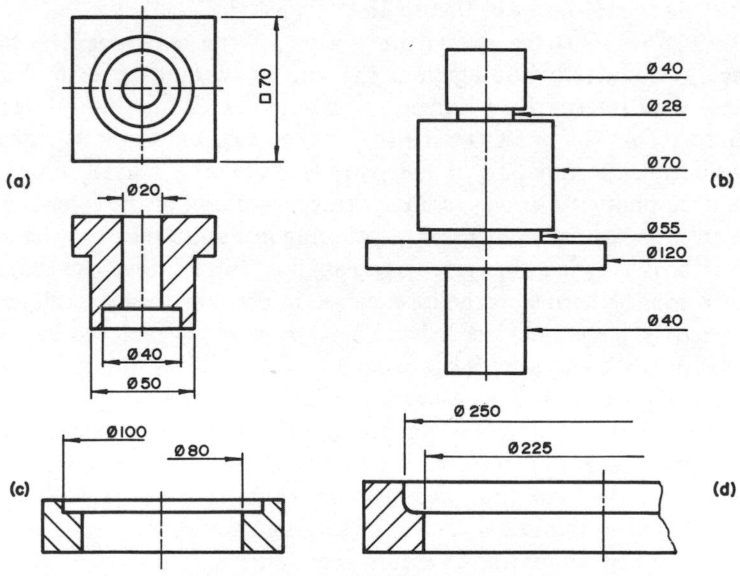

FIG.7   DIMENSIONING DIAMETERS

69

extension line. Such symbols are called machining marks. For a detail which is to be machined all over, a general note ' √ all over' may be used.

With parts which are machined all over, or which are not machined at all, suitable datum faces or lines must be selected to ensure the correct functioning of the part, or to aid manufacture. The pin shown in Figure 15, for example, has its datum face under the head to give the machinist the dimensions he requires directly.

Similarly, the positions of the holes in the drilled plate in the same figure need to be related to the edges AB and BC for functional reasons, so they are located from these edges.

Redundant dimensions

A drawing must carry only those dimensions needed to make the part, and no dimension must appear more than once on a view, or on different views, neither must information be given in two different ways. The reason for this is that dimensions on drawings are sometimes altered when the design is modified, and if they appear in more than one place one of them may be missed. The same dimension will then occur with two different values and cause confusion. Examples of redundant dimensions are shown in Figure 16.

In Figure 16(a) the overall dimension of the component is given twice, once directly and again as the sum of two intermediate dimensions. One intermediate dimension should be omitted, as shown in Figure 16(b). The one to be omitted is the least important, decided by considering the function of the part. Equally, the overall dimension could be omitted if it was less important than any of the intermediate dimensions. Occasionally the dimensioning in Figure 16(c) may be used, when the overall length is given as an auxiliary dimension. The auxiliary dimension is identified by being enclosed in brackets. The overall length of a part is sometimes needed to decide how much material to order, or to cut stock into approximate lengths. An auxiliary dimension is not used in the machining or inspection of a part.

Dimensions of features which are obviously the same should not be repeated. This is illustrated in Figures 16(c) and (d). The corner radii and thickness of the four webs should be given once only. Similarly fillet and blend radii on a casting or forging should be given once, and it is convenient to do this by means of a general note.

To ensure that no redundant dimensions have been given and to check the completeness of the dimensioning, the draughtsman should mentally go through the manufacture of the part he has drawn, at the same time marking each dimension as he uses it. Any omitted dimensions will be apparent as the check proceeds. At the end, any unmarked dimensions will be superfluous.

DIMENSIONING

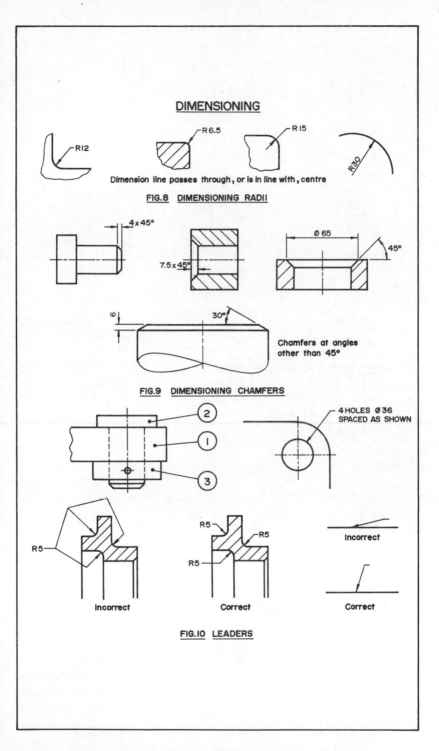

R12

R6.5

R15

R30

Dimension line passes through, or is in line with, centre

FIG.8   DIMENSIONING RADII

4 x 45°

7.5 x 45°

Ø 65

45°

6

30°

Chamfers at angles
other than 45°

FIG.9   DIMENSIONING CHAMFERS

2

1

3

4 HOLES Ø 36
SPACED AS SHOWN

R5

Incorrect

R5
R5
R5

Correct

Incorrect

Correct

Incorrect

Correct

FIG.10   LEADERS

DIMENSIONING

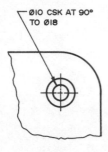

Ø10 CSK AT 90°
TO Ø18

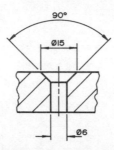

90°
Ø15

Ø6

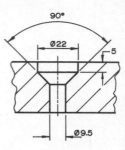

90°
Ø22

5

Ø9.5

FIG.11 DIMENSIONING COUNTERSINKS

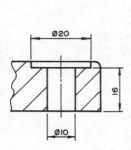

Ø20

16

Ø10

Ø16 C'BORE Ø25 x 6 DEEP

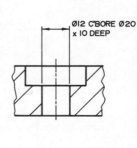

Ø12 C'BORE Ø20
x 10 DEEP

FIG.12 DIMENSIONING COUNTERBORES

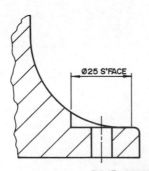

Ø25 S'FACE

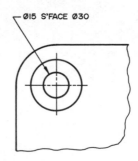

Ø15 S'FACE Ø30

FIG.13 DIMENSIONING SPOTFACES

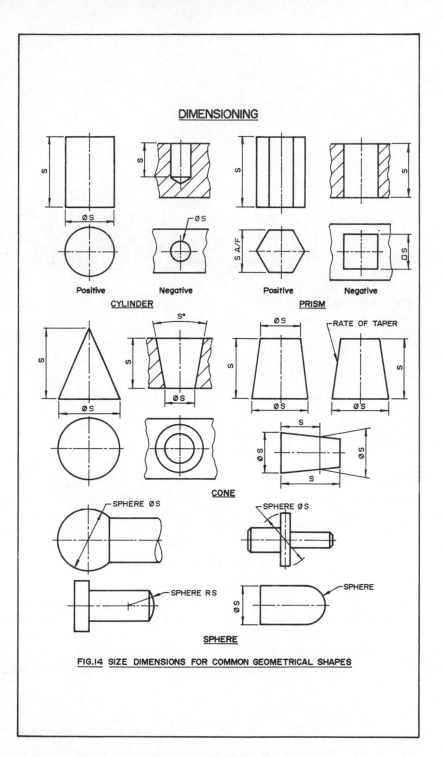

DIMENSIONING

Positive **Negative** **Positive** **Negative**

CYLINDER **PRISM**

CONE

SPHERE

FIG.14 SIZE DIMENSIONS FOR COMMON GEOMETRICAL SHAPES

73

DIMENSIONING

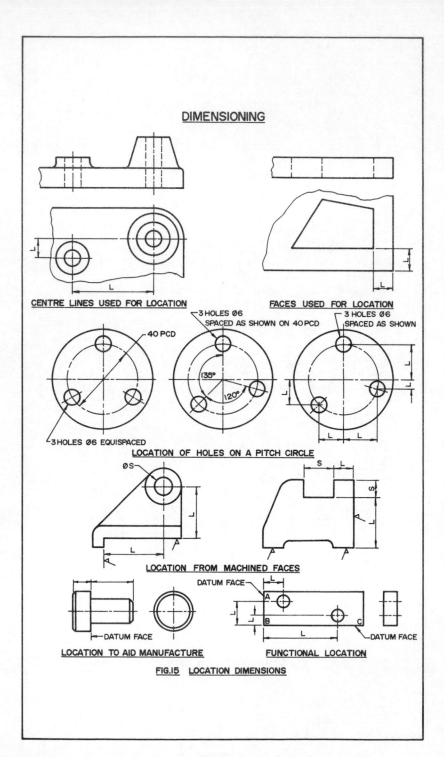

CENTRE LINES USED FOR LOCATION

FACES USED FOR LOCATION

40 PCD

3 HOLES Ø6
SPACED AS SHOWN ON 40 PCD

3 HOLES Ø6
SPACED AS SHOWN

135°

120°

3 HOLES Ø6 EQUISPACED

LOCATION OF HOLES ON A PITCH CIRCLE

ØS

LOCATION FROM MACHINED FACES

DATUM FACE

DATUM FACE

LOCATION TO AID MANUFACTURE

DATUM FACE

A

B

C

DATUM FACE

FUNCTIONAL LOCATION

FIG.15   LOCATION DIMENSIONS

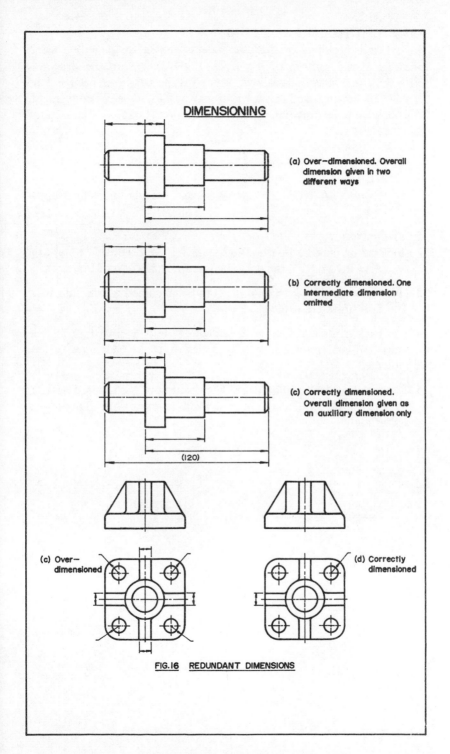

DIMENSIONING

(a) Over—dimensioned. Overall dimension given in two different ways

(b) Correctly dimensioned. One intermediate dimension omitted

(c) Correctly dimensioned. Overall dimension given as an auxiliary dimension only

(120)

(c) Over—dimensioned

(d) Correctly dimensioned

FIG.16   REDUNDANT DIMENSIONS

75

DIMENSIONING PROBLEMS

Make detail drawings of the components shown on page 77, using the scale at the bottom of the page. Fully dimension the drawings following the instructions below. Views on the solutions need not be those given. Sections and conventions may be used as seems convenient. Solutions are to be drawn in the same projection system as that used for the given views.

1 (a) Locate each hole from the right-hand and top edges of the detail.
  (b) Redraw assuming hole A to be the datum and dimension accordingly.

2 (a) Dimension the plate from the edges XY and YZ.
  (b) Redraw making hole A the datum for the other two holes and edges PQ and YZ the datum edges for the group of holes.

3 Dimension the bracket assuming it to be machined where indicated. The four holes are drilled.

4 The back machined face of the flange is the datum surface for the shaft support. Dimension it assuming it to be machined where indicated.

5 The detail is machined all over. The position of the tapered portion relative to the underside of the head is important. Dimension accordingly.

DIMENSIONING PROBLEMS

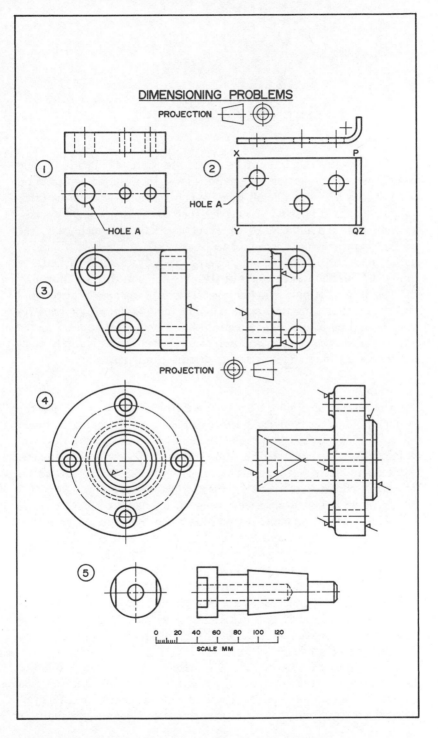

PROJECTION

①

②

HOLE A

HOLE A

X   P

Y   QZ

③

PROJECTION

④

⑤

0 20 40 60 80 100 120
SCALE MM

77

7

SECTIONAL VIEWS

OBJECTS with little interior detail can be represented satisfactorily in orthographic projection by exterior views, the interior construction being shown by hidden detail lines: When the interior detail is more complicated, as in Figure 1 on page 79, then the hidden detail lines may be confusing and difficult to interpret correctly. In such cases the draughtsman imagines the object to be cut by a plane as in Figure 2, and assumes the part of the object between his eye and the plane to be removed. This exposes the interior detail which can then be shown by full lines instead of hidden detail lines. The resulting view is a sectional view or a section. Strictly a sectional view includes all visible lines behind the section plane, while a section shows only what appears on the cutting plane. A section, as opposed to a sectional view, is rarely used, and frequently the two terms are used indiscriminately.

Full sections
The view shown in Figure 2 is called a full sectional view because the cutting plane passes completely through the object. It should be noted that all visible edges behind the plane must be shown or the view will be incomplete, as illustrated in Figure 3. Such a view is meaningless. Hidden detail lines, however, are not shown on a sectional view unless they are needed to describe the object completely. The position of a cutting plane is shown on a view where it appears as a line, and the direction in which the plane is viewed is given by arrows at each end. Letters on the arrows and a title such as 'Section AA' below the sectional view, relate the view to the cutting plane. The cutting plane line is a long thin chain line with a thick long dash at each end. The arrows are placed with their points touching the centre of this thick dash. Note that the other view or views on the drawing show the complete object, unless they also are sectional views. This is because the object is only imagined to be cut by the section plane. These points are illustrated in Figure 2.

The position of the cutting plane is selected by the draughtsman to show the interior of the object to the best advantage. When its position is obvious, for example when it coincides with the centre line of a symmetrical object, it and the title of the sectional view are often omitted.

SECTIONAL VIEWS

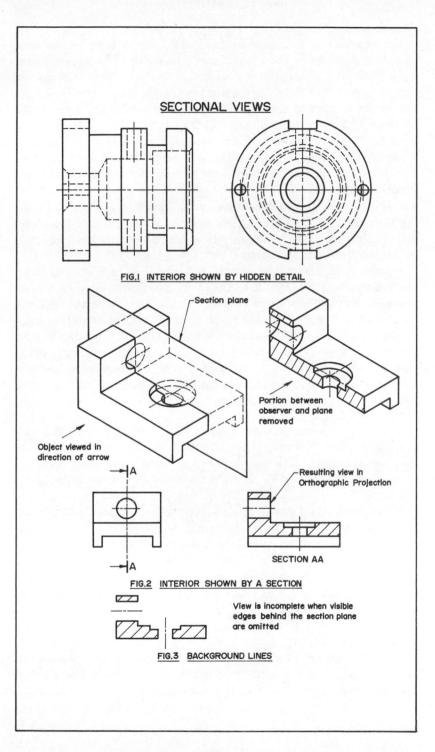

FIG.1 INTERIOR SHOWN BY HIDDEN DETAIL

Section plane

Portion between
observer and plane
removed

Object viewed in
direction of arrow

A

A

Resulting view in
Orthographic Projection

SECTION AA

FIG.2 INTERIOR SHOWN BY A SECTION

View is incomplete when visible
edges behind the section plane
are omitted

FIG.3 BACKGROUND LINES

79

Section lines

A sectional view is distinguished from an outside view by section lines or hatching drawn on the cut surfaces produced by the section plane. Thin lines, inclined at 45°, are used for hatching. They are equally spaced by eye, the spacing being not less than about 4 mm. The larger the area to be hatched the wider the spacing may be, up to a maximum of about 10 mm. For very large areas the hatching may be limited to a zone following the contour of the hatched area as shown in Figure 7. If the meaning of the drawing is clear without hatching a sectional view, it may be omitted. This has been done in Figures 9 and 12 and at other places in this book.

It is important to the appearance of the finished drawing that hatching is carefully drawn. The spacing must be consistent, not too close, and the lines must touch the outlines of the section and be thin. These points are illustrated in Figure 4.

When a single component is sectioned, as in Figure 5, the slope and spacing of the hatching must be the same throughout the view. On an assembly drawing the slope of the hatching must be reversed on adjacent parts. For each component the slope and spacing of the hatching must be the same on all views on the drawing. If more than two parts are in contact, the angle of the hatching may be changed to some angle other than 45°, or the spacing may be varied to avoid the impression that it is crossing outlines. Many elementary mistakes in hatching sectional views will be avoided if it is remembered that hatching can never cross an outline.

Cases sometimes occur when the outlines or axes of a sectional view slope at 45°. If the hatching is also drawn at 45° a misleading impression is given. It should therefore be drawn at some other easily obtained angle, such as 30° or 60°, or drawn horizontally or vertically, as in Figure 6.

When sheet metal parts, gaskets and other thin details appear on sectional views, the areas involved may be too narrow for hatching. Instead of drawing the material thickness out of scale, the section may be blacked in as shown in Figure 8. If two or more thin details are joined on a sectional assembly drawing, spaces should be left between them or the drawing will be difficult to read. This has been done on the built-up girder in Figure 8.

It was customary in the past to indicate different materials on a sectional view by different types of hatching, such as broken lines and alternate broken and full lines. Nowadays, with the large number of materials in common use, this practice has been entirely discontinued and the hatching described above is used exclusively.

SECTIONAL VIEWS

| Uneven spacing | Spacing too close | Not touching outlines | Too thick | Correct |

FIG.4 FAULTS IN SECTION LINING

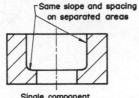 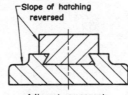

┌Same slope and spacing on separated areas   ┌Slope of hatching reversed

Single component   Adjacent components

FIG.5 SECTION LINING OF ADJACENT COMPONENTS

  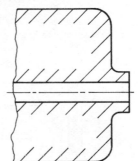

Incorrect. Hatching parallel to outlines or axes

Correct

FIG.6 ALTERATION OF SLOPE OF HATCHING   FIG.7 HATCHING OF LARGE AREAS

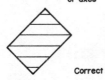 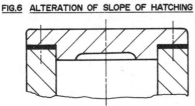 

Gasket   Built-up girder

FIG.8 SECTIONING OF THIN DETAILS

SECTIONAL VIEWS

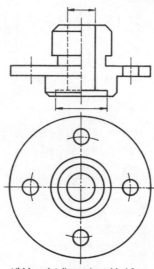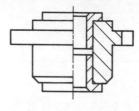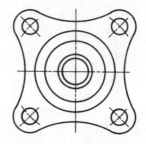

Hidden detail may be added for
dimensioning

Hidden detail omitted for an
assembly

FIG. 9   HALF SECTIONS

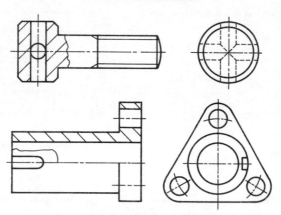

FIG. 10   LOCAL OR BROKEN-OUT SECTIONS

SECTIONAL VIEWS

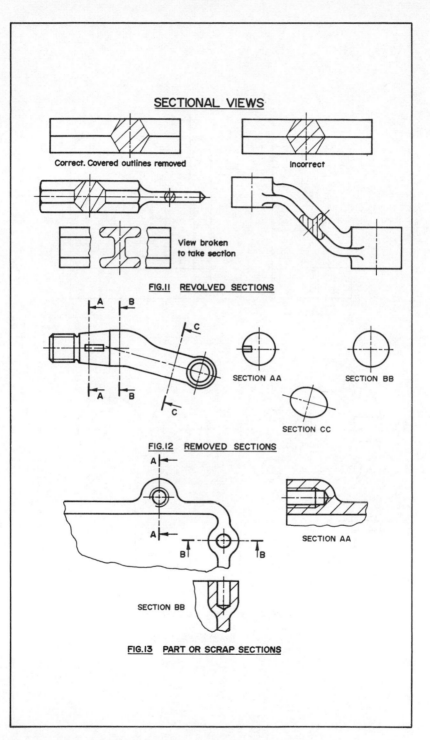

Correct. Covered outlines removed

Incorrect

View broken to take section

FIG.11 REVOLVED SECTIONS

SECTION AA

SECTION BB

SECTION CC

FIG.12 REMOVED SECTIONS

SECTION AA

SECTION BB

FIG.13 PART OR SCRAP SECTIONS

83

SECTIONAL VIEWS

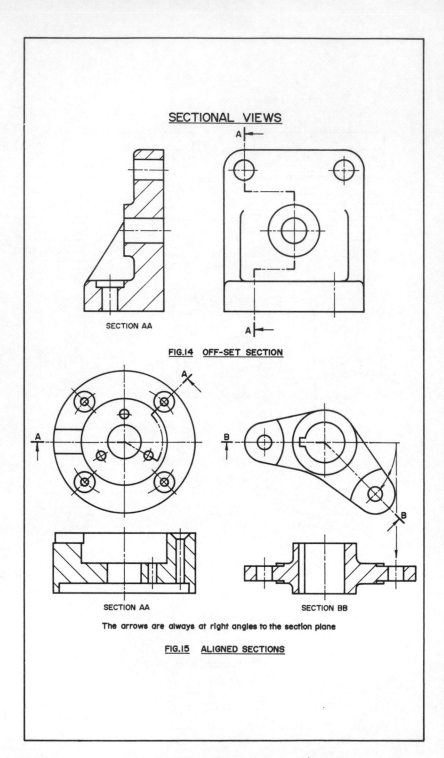

SECTION AA

FIG.14   OFF-SET SECTION

SECTION AA

SECTION BB

The arrows are always at right angles to the section plane

FIG.15   ALIGNED SECTIONS

84

SECTIONAL VIEWS

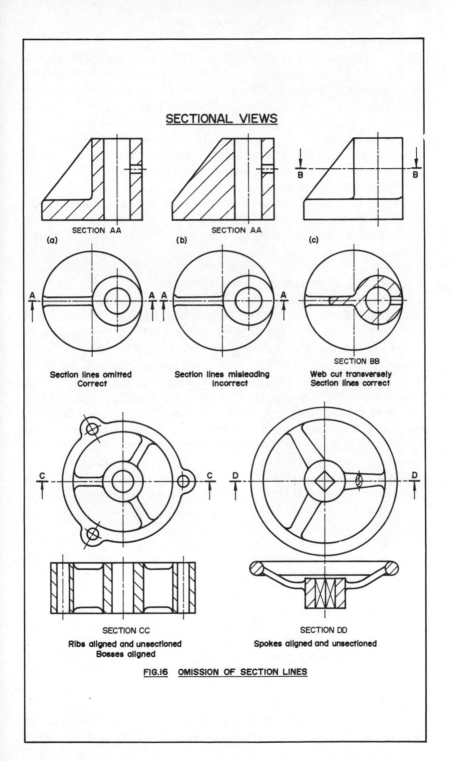

SECTION AA

(a)

SECTION AA

(b)

(c)

Section lines omitted
Correct

Section lines misleading
Incorrect

SECTION BB

Web cut transversely
Section lines correct

SECTION CC

Ribs aligned and unsectioned
Bosses aligned

SECTION DD

Spokes aligned and unsectioned

FIG.16   OMISSION OF SECTION LINES

85

Half sections

Symmetrical objects may be shown to advantage by half sections, that is with one half drawn in section and the other as an outside view. Thus two views are combined in one with a consequent saving of draughting time and space. They have the disadvantage that the dimensioning of the internal features is often difficult without using hidden detail lines, and for this reason they are more often used for assembly drawings. Few dimensions, if any, are required on assembly drawings, and hidden detail can be omitted from the unsectioned half of the view. Examples of half sections are given in Figure 9.

The halves of a half section should be separated by a centre line and not by an outline. This is because the object is only imagined to be cut by the section plane, and there is, therefore, no visible line at the centre of the view.

Local or broken-out sections

Occasionally a part of a view in section is all that is needed to show the internal details of an object, and sometimes a whole view cannot be sectioned because it would conceal external features which must be shown. In such cases a local or broken-out section is useful.

The cutting plane is imagined to pass part way through the object and the piece in front of the plane is then assumed to be broken away, leaving an irregular boundary. Note that this boundary is a thin line. In Figure 10 the adjusting screw is solid except for the two holes in the head, and these are shown by a local section through the head. In the other detail a complete half section would have removed half of the keyway, producing a view which might have been misleading. The keyway is preserved in its entirety by stopping the half section with a break line. Other examples of local sections are included in Figure 17.

Revolved sections

These sections are drawn directly on an outside view and are used to show the local cross-section of an arm, rib or similar feature. The cutting plane is assumed to pass through the part at right angles to the axis and is then revolved into the plane of the paper. Examples of revolved sections are given in Figure 11. Note that the outline of the section is a thin line, and that any outlines on the main view which are covered by the section are not shown. Revolved sections are often placed in the gap between the parts of an interrupted view, but this practice is not illustrated in BS 308:1972.

Removed sections

These are similar to revolved sections but instead of being drawn on the outside view they are removed to another part of the drawing. Their outlines are the normal thick line. They are used in place of revolved sections when lack of space prevents the section being drawn or dimensioned on the outside view. They are sometimes drawn to a larger scale than the remainder of the drawing, are not subject to the strict rules of projection regarding their position on the drawing, and, if numerous, are occasionally grouped together on a separate sheet. For these reasons it is essential that the position of the cutting plane is stated and the section labelled to relate it to the cutting plane. Removed sections are illustrated in Figure 12. They are examples of true sections as distinct from sectional views, since features behind the cutting plane are omitted.

Part or scrap sections

It is necessary occasionally to show part of an object in section to describe a small detail of the construction, but the views may be unsuitable for the use of a broken-out section. In such a case a part or scrap section may be used. This is similar to a broken-out section but is drawn away from the outside view. The scrap section has an irregular boundary, which is thin, and frequently background features are omitted. Examples are shown in Figure 13. Scrap sections, like removed sections, are often drawn in any convenient position on the drawing, and should, therefore, carry a title.

Off-set sections

On a full section the cutting plane need not pass straight through an object, but may be off-set as required to include features which are not in a straight line. This is illustrated in Figure 14. The position of the cutting plane must be shown in a view where it appears as a line, and the resulting view should carry a title. Where the plane changes direction, thick dashes are used and these dashes touch to form a right angle. Note that at the places where the plane changes direction on the sectional view, no lines appear. There are two reasons for this. First, the cutting of the detail by the plane is imaginary, and secondly such lines would be crossed by the hatching. This would violate the rule that hatching can never cross outlines.

Aligned sections

These are used when sectioning parts whose features lie on radial lines, such as those shown in Figure 15. The cutting plane generally coincides with a principal centre line and one or more radial centre lines. The section is drawn with the features on the radial centre lines aligned with the principal centre line. By this means, awkward projections are avoided and the features are shown at their true distances from the centre of the part. As with off-set sections, where the cutting plane changes direction it is drawn as a thick line, and these thick lines touch. Other examples of aligned sections are shown in Figure 16.

Alignment is also used on outside views to avoid tedious projections. For example, if the section of the bell crank lever in Figure 15 were drawn as an outside view, it could still be aligned to make the drawing easier.

Features from which hatching is omitted

Whenever hatching would result in a misleading effect, it should be omitted. For example, consider the casting shown in Figure 16 at (a), (b) and (c). Section AA passes longitudinally through a triangular web whose purpose is to strengthen the component. If the section lines are drawn on the web as at (b), the impression of a partly conical shape is given. This impression is incorrect so the hatching is omitted and (a) is the correct representation. Note that the web is bounded by an outline where it runs into the base and circular boss. It is a common error for beginners to omit these outlines or to use hidden detail lines. If the web or rib is cut transversely by the cutting plane as at (c), it must always carry the hatching.

Further examples of the omission of hatching are given on sections CC and DD in Figure 16. Here the ribs and spokes are not hatched and to avoid tedious projections they are aligned on to the plane of the section. Gear teeth are treated in a similar way.

Details not sectioned

In addition to the examples given above of the omission of hatching, some machine details are not sectioned on assembly drawings when they are cut longitudinally by the section plane. These details include nuts, bolts, screws, studs, rivets, solid shafts and small solid cylindrical parts, keys, cotters, split and taper pins and balls and rollers in bearings. These parts are not sectioned because they have no internal features, and also because they are more easily recognized by their outside views than by a section. Figure 17 shows two small assemblies to which this principle has been applied. However, if these details are cut transversely, resulting in a circular sectional view, they are hatched in the normal way.

SECTIONAL VIEWS

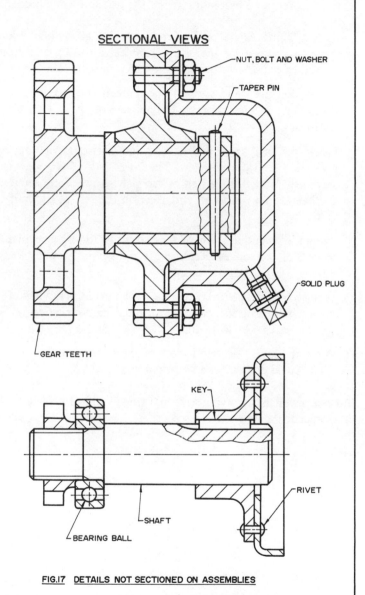

NUT, BOLT AND WASHER

TAPER PIN

SOLID PLUG

GEAR TEETH

KEY

RIVET

SHAFT

BEARING BALL

FIG.17 DETAILS NOT SECTIONED ON ASSEMBLIES

89

SECTIONAL VIEW PROBLEMS

Draw the following problems full size using First or Third Angle projections as required by the question. Do not show any hidden edges.

1 Using Third Angle projection draw the given plan view of the bearing and replace the given front elevation by a sectional front elevation on AA. Add a sectional end view on BB.

2 Draw the given end view of the flanged coupling and replace the given outside elevation by a half section AA. Use First Angle projection.

3 Draw in Third Angle projection a half sectional view replacing the given left-hand outside view. The lower half of this view is to be in section. On the outside half of the view show a local section around the M8 tapping. Add an end view on the left of the half sectional view.

4 Copy the given half view of the assembly adding a revolved section on one of the spokes. Replace the right-hand view by a section on AA. Use First Angle projection.

5 Using Third Angle projection draw the given plan view of the component and project from it a sectional front elevation on AA. Add a sectional end view on BB.

6 Draw the given left-hand view of the cover and project from it in First Angle projection a sectional view on AA.

Further problems involving sectional views will be found in the Machine Drawings at the end of the book.

SECTIONAL VIEW PROBLEMS

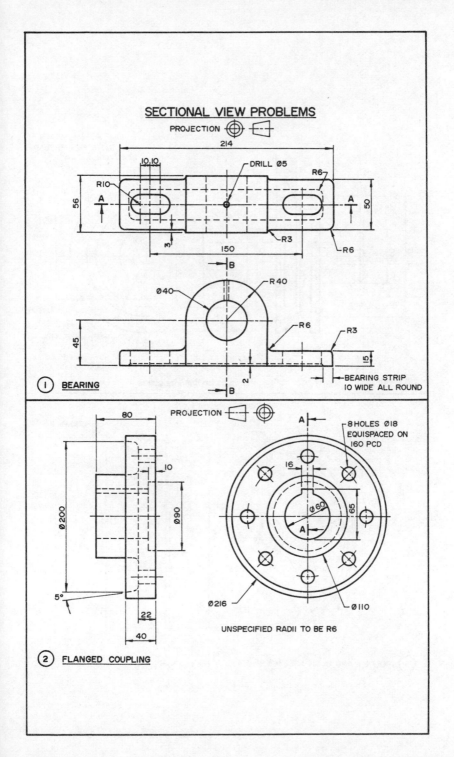

PROJECTION ⊕ ◁

214

10,10

DRILL Ø5

R6

RIO

A

A

56

50

B

Ø40

R40

R6

R3

45

150

R6

R3

15

2

① BEARING

B

BEARING STRIP
10 WIDE ALL ROUND

PROJECTION ◁ ⊕

80

A

8 HOLES Ø18
EQUISPACED ON
160 PCD

10

16

Ø200

Ø90

Ø60

65

A

5°

22

40

Ø216

Ø110

UNSPECIFIED RADII TO BE R6

② FLANGED COUPLING

91

SECTIONAL VIEW PROBLEMS

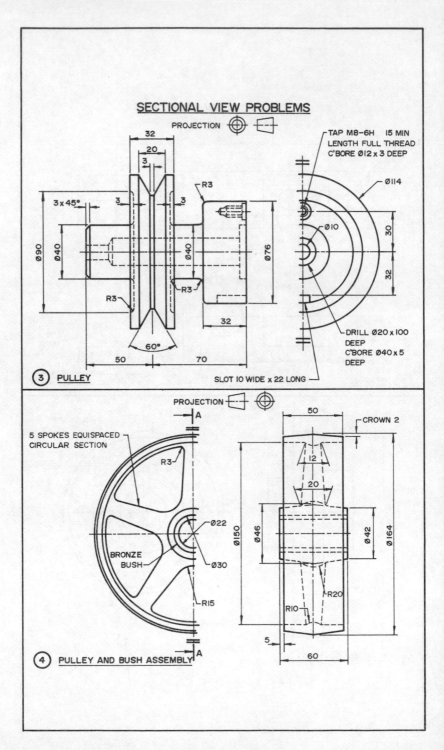

PROJECTION

TAP M8-6H 15 MIN
LENGTH FULL THREAD
C'BORE Ø12 x 3 DEEP

32
20
3
3 x 45°
3
R3
R3
R3
60°
3
Ø90
Ø40
Ø40
Ø76
50
70
32

Ø114
Ø10
30
32

DRILL Ø20 x 100 DEEP
C'BORE Ø40 x 5 DEEP

SLOT 10 WIDE x 22 LONG

(3) PULLEY

PROJECTION

5 SPOKES EQUISPACED
CIRCULAR SECTION

R3
Ø22
BRONZE BUSH
Ø30
R15

50
CROWN 2
12
20
Ø150
Ø46
Ø42
Ø164
R20
R10
5
60

(4) PULLEY AND BUSH ASSEMBLY

92

SECTIONAL VIEW PROBLEMS

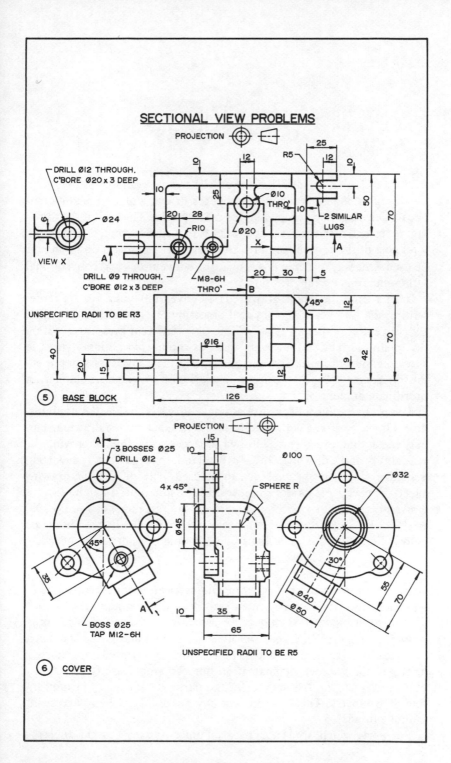

PROJECTION

DRILL Ø12 THROUGH.
C'BORE Ø20 x 3 DEEP

Ø24

VIEW X

R5

25
12
10

Ø10 THRO'

2 SIMILAR LUGS

50

70

20 28

R10

Ø20

M8-6H THRO'

DRILL Ø9 THROUGH.
C'BORE Ø12 x 3 DEEP

X

A

B

20 30 5

UNSPECIFIED RADII TO BE R3

45°

Ø16

70

40

20

15

42

9

B

126

⑤ BASE BLOCK

PROJECTION

A
3 BOSSES Ø25
DRILL Ø12

15
10

4 x 45°

Ø45

SPHERE R

Ø100

Ø32

45°

30°

35

A

10 35

65

Ø40

Ø50

55

70

BOSS Ø25
TAP M12-6H

UNSPECIFIED RADII TO BE R5

⑥ COVER

93

8

CONVENTIONS

PRODUCING the drawing for a very simple component may take the draughtsman several hours, whilst the drawing of a large and complicated casting may occupy him for many weeks. The cost of the drawing, represented by the time the draughtsman spends on it, is part of the total cost of the component, and any means of keeping it down should be used. One way of doing this is by using conventions and conventional representations on the drawing.

In one sense the whole drawing is a convention, since we are representing the surfaces of the physical object by lines on a sheet of paper. By convention, different types of line have different meanings. Within this main convention are others, such as the use of sectional views to show the interior features of the object, and the conventional representation of screw threads, the accurate projection of which would take an inordinate amount of time.

The primary object of the use of conventions is to save the draughtsman's time, but some conventions also save space on the drawing. This may mean that a view can be drawn to larger scale, thus improving the readability of the drawing. On the other hand, space saved may mean that a smaller size of drawing sheet can be used. The smaller the drawing sheet, the cheaper is its storage, transmission and reproduction. Various conventions to achieve both these objects are recommended in BS 308: Part 1:1972, and those which occur most frequently are discussed below. For a full treatment the student should consult the Standard.

Symmetry
Many engineering components are symmetrical about a centre line or axis and can often be represented satisfactorily by a half view. This is illustrated in Figure 1(a) on page 97. To show that a half view has been drawn, two short, thick, parallel lines are drawn across the symmetry demarcation line at each end. These symmetry symbols are at right angles to the symmetry demarcation line. To emphasize further that a half view is shown, the outlines of the part extend slightly beyond the line of symmetry. This is to prevent any possibility of half parts being manufactured.

For parts which are symmetrical about two axes at right angles, a quarter view may be used. An example is shown in Figure 1(b).

Sometimes components which are basically symmetrical have asymmetrical features. The above convention may still be used in such cases, provided that the half view which shows the asymmetrical features is drawn, and that the asymmetrical features are identified by a note. Figure 1(c) is an example.

It may be sufficient to show an adequate segment of a circular part, as illustrated in Figure 1(d). However, all uses of this convention should be considered carefully to avoid any ambiguities or loss of understanding of the drawing. A drawing which is not clear will waste time and may result in the production of scrap.

Enlarged part views
On occasion, components carry features which, in the general scale of the drawing, are too small to be dimensioned clearly. Such features may be enclosed in a thin circle from which a leader is drawn to an enlarged part view. This part view is then used for the dimensions. Figure 2 illustrates this convention.

Repetitive information
On some drawings identical parts or features appear many times. Repeated illustrations of them may be avoided by drawing one, and indicating the positions of the others by their centre lines, as shown in Figure 3(a).

If the repeated parts are required in a regular pattern, only the number necessary to establish the pattern need be shown, using their centre lines. The rest of the information should be given in a note. This is illustrated in Figure 3(b), which shows a riveted butt joint.

Figure 3(c) shows a component which has twelve identical holes. Adjacent to two pairs of these holes are single features. The holes on each side of the single features should be shown in full.

Conventional representation of common features
Figure 4 shows the conventions for those common features which occur most often, and are selected from BS 308: Part 1:1972. The conventions for screw threads are given in Chapter 9.

Long parts of constant cross-section may be shown by interrupted views. A piece from the centre of the part is imagined to be removed and the ends of the view are brought together. Three break lines for different cross-sections are used as shown. These break lines are thin. Sometimes a revolved section is drawn in the gap in the view.

The circumferential surface of the heads of adjusting screws is often roughened by knurling to provide a better finger grip. Two types, straight and diamond knurling, are in use and the conventional representations use thin lines. The spacing of the lines should be quite wide to avoid them closing up on reduced size prints. Straight knurling is often used on the heads of hexagon socket cap screws.

A square end on a shaft, provided to give a spanner grip or to mount a hand wheel, is represented by thin diagonals drawn on the flat face. A similar convention is used for ball and roller bearings, the balls, rollers and cages being omitted.

The helices on a cylindrical compression spring should not be projected but drawn as straight lines. The pitch of the coils need not be exact and only two or three coils should be shown at each end of the spring. The remaining coils are shown by thin, chain lines. These are sometimes called ditto lines. The end coils are closed and ground square with the axis to provide a flat surface at each end of the spring, and this is shown on the conventional representation. For diagrams and schematic drawings the representation may be simplified to a single line as shown.

Teeth on the circular view of a gear may be omitted and shown conventionally by a thick circle representing the tips of the teeth, and a thin, chain circle representing the pitch circle. No circle is shown through the roots. On a sectional view of a gear, hatching is omitted from the teeth, even though the cutting plane passes through them. The pitch line is shown by a thin, chain line.

For a rack, one or two teeth are drawn at each end and the remainder shown by a thick line representing the tips and a thin line representing the roots. The pitch line is not shown. The dimensions indicate how the teeth are located on the blank.

CONVENTIONS

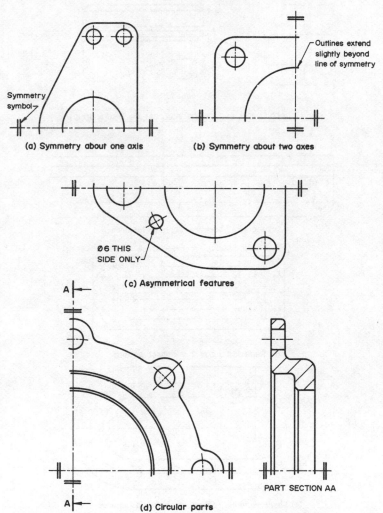

Symmetry symbol

(a) Symmetry about one axis

Outlines extend slightly beyond line of symmetry

(b) Symmetry about two axes

Ø6 THIS SIDE ONLY

(c) Asymmetrical features

A

A

PART SECTION AA

(d) Circular parts

Do not use this convention if loss of understanding of the drawing would result

FIG.I SYMMETRY

CONVENTIONS

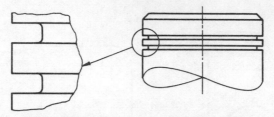

FIG.2 PART VIEW ENLARGED FOR CLEAR DIMENSIONING

(a) Repeated identical parts or features

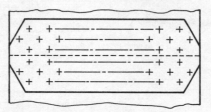

(b) Repeated parts in a regular pattern

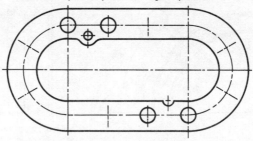

(c) Single features adjacent to repeated features

FIG.3 REPETITIVE INFORMATION

CONVENTIONS

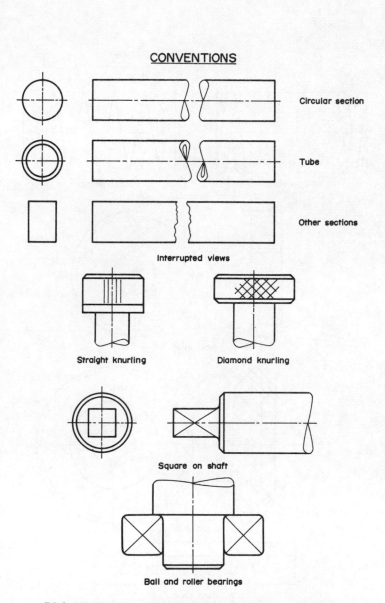

Circular section

Tube

Other sections

Interrupted views

Straight knurling   Diamond knurling

Square on shaft

Ball and roller bearings

FIG. 4 CONVENTIONAL REPRESENTATION OF COMMON FEATURES

CONVENTIONS

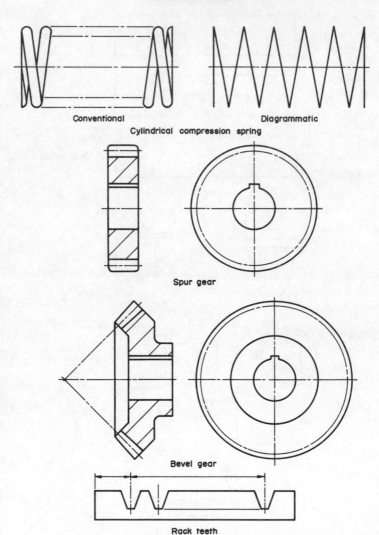

Conventional   Diagrammatic

Cylindrical compression spring

Spur gear

Bevel gear

Rack teeth

FIG.4 (CONT'D)   CONVENTIONAL REPRESENTATION OF COMMON FEATURES

9

SCREW THREADS

A SCREW thread is a helical groove which is cut, rolled or sometimes cast on a cylinder or in a cylindrical hole. These are parallel threads, that on the cylinder, or screw, being external, and that in the hole, or nut, being internal. Tapered threads are formed on a cone or in a conical hole. The form of the groove varies with different threads, depending on the use to which it is to be put. Threads may be right- or left-handed and single- or multi-start, as explained in Chapter 4 in the section on helices.

Screw thread terms
These are illustrated in Figure 1 on page 103.

Pitch
The pitch of a thread is the distance from a point on one thread form to the corresponding point on the next thread form, measured parallel to the axis in an axial plane. The lead of a thread is the axial movement of the screw in one revolution. For single-start threads pitch and lead are the same. For multi-start threads the pitch is the distance from a point on one start to the corresponding point on the next start. So for a two-start thread, pitch is half the lead; for a three-start thread, pitch is a third of the lead, and so on.

Crest
The crest is the prominent part of the thread, whether it be external or internal.

Ròot
The root is the bottom of the groove between adjacent thread forms.

Flanks
The flanks of a thread are the straight sides which connect the crest and root.

Thread angle
This is the angle between the flanks measured in an axial plane.

Major diameter
The diameter over the crests of an external thread, or the roots of an internal thread, measured at right angles to the axis. Sometimes called the full or outside diameter.

Minor diameter
The diameter at the roots of an external thread, or the crests of an internal thread, measured at right angles to the axis. It is also known as the core diameter.

Effective diameter
The diameter of a cylinder co-axial with the thread, which cuts the flanks of a thread form in two points such that the distance between them is half the pitch. Also called the pitch diameter.

Thread depth
This is half the difference between the major and minor diameters.

Conventional representation of screw threads
The correct projection of a screw thread is tedious and takes a considerable time, so threads are shown conventionally on engineering drawings. The conventions, taken from BS 308: Part 1:1972, are shown in Figure 2.

On the longitudinal view of an external thread, the major diameter is shown by a pair of thick lines, and the minor diameter by a pair of thin lines. The end of the full thread, that is, the point at which the root of the thread ceases to be fully formed, is drawn as a thick line. Beyond the full thread is the thread run-out, consisting of incompletely formed threads produced by the lead-in chamfer on the die. The run-out is represented by thin lines at 30° to the major diameter. On the circular view the major diameter appears as a complete thick circle and the minor diameter as a thin circle with a gap.

For an internal thread in section the minor diameter drilling is drawn in thick lines, as is the line representing the end of the full thread. The major diameter and thread run-out are shown by thin lines. The hatching crosses the major diameter and terminates on the minor diameter. If an internal thread is shown in hidden detail, all lines are thin short dashes. The circular view shows the minor diameter as a complete thick circle and the major diameter as a thin circle with a gap.

On a sectional thread assembly the screw is shown as an outside view so its threads cover the threads in the tapping. Below the screw the internal threads appear as described in the previous paragraph, and the

SCREW THREADS

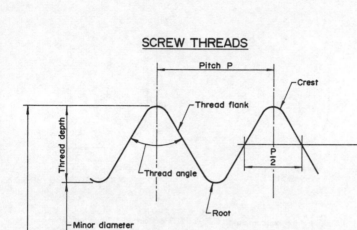

FIG.I   SCREW THREAD TERMS

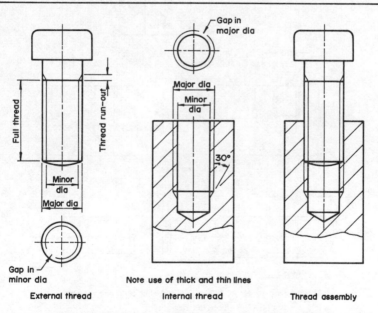

FIG.2   CONVENTIONAL REPRESENTATION OF SCREW THREADS

SCREW THREADS

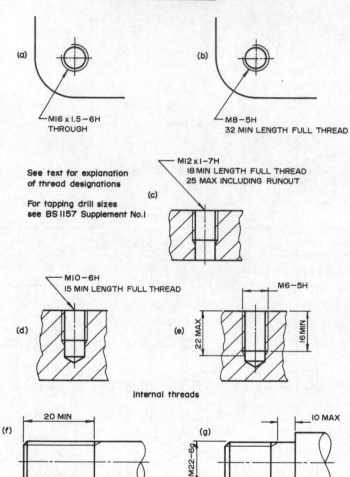

(a) M16 x 1.5 – 6H
THROUGH

(b) M8 – 5H
32 MIN LENGTH FULL THREAD

See text for explanation
of thread designations

For tapping drill sizes
see BS 1157 Supplement No.1

(c) M12 x 1 – 7H
18 MIN LENGTH FULL THREAD
25 MAX INCLUDING RUNOUT

(d) M10 – 6H
15 MIN LENGTH FULL THREAD

(e) M6 – 5H
22 MAX
16 MIN

Internal threads

(f) 20 MIN
28 MAX

(g) 10 MAX
M22 – 6g

Dimensioning to end of full thread and limit of thread run-out

External threads

FIG.3   DIMENSIONING OF SCREW THREADS

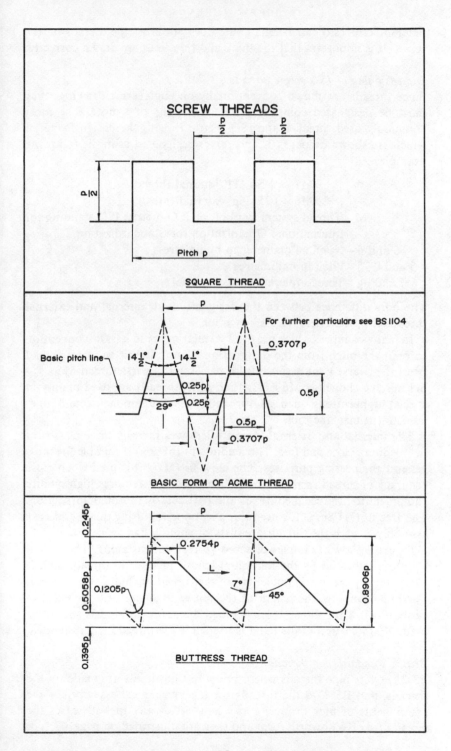

SCREW THREADS

SQUARE THREAD

BASIC FORM OF ACME THREAD

For further particulars see BS 1104

BUTTRESS THREAD

hatching crosses them. The hatching does not cross the threads on the screw. It is important that the thick and thin lines are drawn correctly.

Dimensioning of ISO screw threads
Since threads are shown conventionally on engineering drawings they must be designated completely by dimensions or a note. The most commonly used thread is the ISO metric thread, the design forms of which are shown on page 107 and this is designated as in the following examples.

M16 × 1·5 – 6H internal thread
M6 × 0·75 – 6g external thread
M – Thread system symbol for ISO metric, ISO standing for International Organization for Standardization.
16 and 6 – Nominal diameter in millimetres
1·5 and 0·75 – Pitch in millimetres
6H and 6g – Thread tolerance class symbol

The only difference between the designations of internal and external threads is in the tolerance class symbol.

In many countries which use the ISO metric thread it is the convention to omit the pitch from the designation. If no pitch is shown a thread from the coarse pitch series is implied. Thus, a coarse thread M6 × 1 – 6H may be shown as M6 – 6H. Fine pitch threads are used mainly in special applications such as for thin-walled components and fine-adjustment machine tools.

The internal and external thread tolerances provide three classes of fit: medium, close and free. The medium fit (6H/6g) is suitable for most general engineering purposes. The close fit (5H/4h) is used when close accuracy of thread form and pitch is needed, but requires high quality equipment to produce the threads and particularly thorough inspection. The free fit (7H/8g) is for use in applications requiring quick and easy assembly, even if the threads are dirty or slightly damaged.

Figure 3 shows examples of screw thread dimensioning. The thread length is indicated by the minimum length of the full thread, and, if necessary, by the maximum length of the thread, including the run-out. For blind tappings the length of the minor diameter drilling will also be required. If the thread is left-handed the abbreviation LH should be used. No indication of the hand is needed if the thread is right-handed.

British standard pipe threads
BS 21 covers pipe threads where pressure-tight joints are made on the threads, and BS 2779 the threads for the mechanical assembly of the components of pipe fittings, cocks and valves etc. In both cases the threads are of Whitworth form and they are illustrated on page 107. The

SCREW THREADS

For further particulars see BS 3643

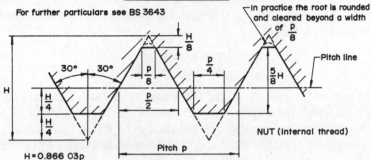

In practice the root is rounded and cleared beyond a width of $\frac{p}{8}$

Pitch line

NUT (Internal thread)

$H = 0.866\ 03p$

ISO METRIC SCREW THREAD

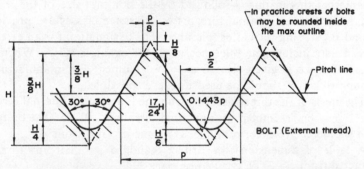

In practice crests of bolts may be rounded inside the max outline

Pitch line

BOLT (External thread)

$H = 0.960\ 491p$
$h = 0.640\ 327p$
$r = 0.137\ 329p$

$H = 0.960\ 237p$ $h = 0.640\ 327p$
$r = 0.137\ 278p$

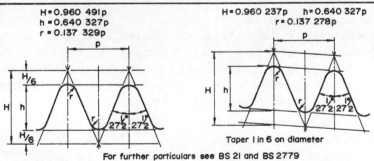

Taper 1 in 6 on diameter

For further particulars see BS 21 and BS 2779

BS PARALLEL PIPE THREAD   BS TAPERED PIPE THREAD

Whitworth thread, designed by Sir Joseph Whitworth over a century ago, was the first British thread to be standardized.

For joints made pressure-tight by the mating of the threads, taper external threads are used with either taper or parallel internal threads, since parallel external threads are not suitable as jointing threads. British Standard taper pipe threads are designated by the letter R for an external thread, or R_c for an internal thread, followed by the nominal size of the thread. For example,

$$\text{Internal taper } R_c\tfrac{1}{2}$$
$$\text{External taper } R\tfrac{1}{2}$$

The nominal size of the thread is the internal diameter in inches of the pipe on which the external thread is cut.

British Standard parallel internal threads to BS 21 are designated by the letters R_p, together with the nominal size of the thread. For example, $R_p\tfrac{1}{2}$.

Fastening pipe threads to BS 2779 are designated by the letter G. For an internal thread this is followed by the nominal size of the thread thus: $G\tfrac{1}{2}$. For an external thread two tolerance classes are provided, A and B, A being the closer tolerance. The designation for an external thread may include the tolerance class, for example, $G\tfrac{1}{2}A$. Where no class reference is given class B is to be assumed. For manufacturing economy class B should be used whenever possible.

The crests of the threads may be truncated, except on internal threads which may be assembled with external threads to BS 21. The designation should then be, for example, $G\tfrac{1}{2}$ trunc or $G\tfrac{1}{2}B$ trunc.

A table of dimensions for British Standard parallel pipe threads, extracted from BS 2779, is given on page 224.

Power transmission threads

Screw threads have three applications. To fasten parts together; to adjust the positions of parts; and to transmit power. The ISO metric thread, being of vee form, is suitable for the first two applications. For power transmission purposes the square, Acme and buttress threads are used, and their forms are illustrated on page 105.

Square thread

Since the flanks of vee threads are inclined to the axis, bursting forces are set up in the nut. The flanks of a square thread are normal to the axis and so these forces are not present. This makes the square thread theoretically ideal for transmitting power. However, the right-angled flanks may cause machining difficulties when the thread is cut in a lathe. The square thread is used on the operating spindles of valves, on jacks, to move the cross slide of a lathe, and similar machine tool applications. It has not been standardized.

Acme thread

This adaptation of the square thread is used where the nut has to be disengaged from the screw, as with the split nut on a lathe lead screw. It is machined more easily than the square thread but the inclined flanks give rise to similar bursting forces in the nut as do vee threads.

Buttress thread

A modern form of this thread is shown in which the front face against which the load is applied is inclined at 7° to the vertical. This face is sometimes vertical. The thread is used to transmit power in one direction only, indicated by the arrow L in the diagram, as in the screw of a vice or press, and the breech blocks of large guns.

10

SCREW FASTENERS

THE commonest form for nuts and bolt heads is hexagonal with a chamfer at the top of the bolt head and at each end of the nut to remove the sharp corners. The hexagonal form is used because it takes up less space, and is therefore lighter, than a square having the same distance across flats. Also one-sixth of a turn brings it into a similar position for a spanner, whereas a square needs a quarter of a turn. An octagonal nut would need only an eighth of a turn, but the flats would be smaller making it easier for a spanner to slip round the corners.

BS 3692 gives the proportions of ISO metric precision hexagon bolts, screws and nuts. A table of the leading dimensions, extracted from this Standard, is given on page 222. Where clearances between nuts and bolt heads and adjacent parts are small, these dimensions should be used so that an accurate idea of the clearances is given. In all other cases, drawing proportions, based on the major diameter of the thread, d, are of sufficient accuracy.

The stages in drawing three views of a hexagon nut using these drawing proportions are set out on pages 111 and 112. The 30° chamfer appears in the hexagonal view as a circle, commonly called the chamfer circle, having the same diameter as the distance across flats. In the other views the chamfers produce hyperbolas on the flats, and these are represented conventionally by circular arcs. If only one elevation of the nut is required on a drawing, that which shows three faces of the hexagon should be used since it is a more characteristic view than the one showing two faces. On the view showing two faces it should be noted that the corners of the view are square and are not removed by chamfers. See stage 7 on page 112.

The head of a hexagon bolt or screw, illustrated on page 113, has the same drawing proportions as a hexagon nut, except that the thickness is reduced from 0.8d to 0.7d. Two types of head are made: washer faced, having a circular face of diameter 1.5d formed on the underside; and full bearing, where the underside is plain. Washer faced heads are made in sizes M3 to M48 only. The nominal length l of a hexagon bolt or screw does not include the thickness of the head. A table of standard nominal lengths from 5 mm to 500 mm is given in BS 3692. The thread length b is calculated from the formulae given on page 113. The only

SCREW FASTENERS

DRAWING HEXAGON NUTS

STAGE 1

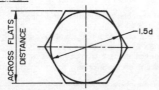

Draw the chamfer circle with diameter 1.5d where d is the major diameter of the thread.
Construct a hexagon round the circle using a 60° set square.

STAGE 2

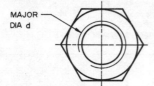

MAJOR DIA d

Complete the view by drawing the two thread circles representing the major and minor diameters.
For a nut the major diameter circle is shown with a break.
The minor diameter circle is broken for a bolt.

STAGE 3

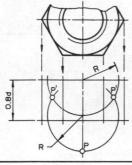

Project the corners of the hexagon to to start the second view.
Make the thickness of the nut 0.8d.
With radius R draw the circle arcs which locate centres P and P'.
R is half the distance across flats.
For a thin nut use 0.6d as thickness.

STAGE 4

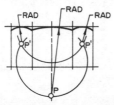

Draw in the chamfer curves using the radii and centres shown.

111

SCREW FASTENERS

DRAWING HEXAGON NUTS

STAGE 5

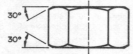

Draw the chamfer curves on the second face.
Add 30° chamfers tangential to radii on outside faces and line in the view.
Note that the 30° chamfers are only apparent on large nuts.

STAGE 6

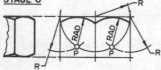

Project the top and bottom faces to start the third view.
Make the width of this view equal to the across flats distance.
With radius R draw the circle arcs which locate centres P.
Draw chamfer curves as shown.

STAGE 7

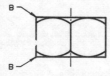

Draw the chamfer curves on the second face.
Line in the view noting the sharp corners at B.

STAGE 8

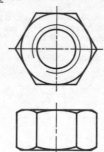

COMPLETED VIEWS

The across flats and thickness dimensions are approximate drawing sizes only. For exact dimensions see BS 3692.

SCREW FASTENERS

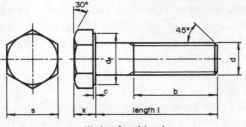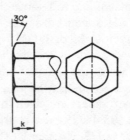

Washer faced head

Full bearing head

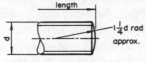

Rounded end

Rolled thread end

Drawing proportions s = 1.5d
k = 0.7d
d_f = 1.5d
c = 1mm

For exact dimensions see
BS 3692

Thread lengths b 2d + 6mm for lengths l up to 125mm
2d + 12mm from over 125mm to 200mm
2d + 25mm over 200mm

Screws are threaded to within two or three thread
pitches of the underside of the head

HEXAGON BOLTS AND SCREWS

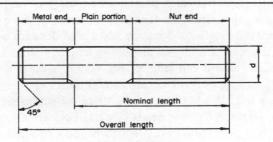

Metal end length d or 1.5d. Includes the runout threads
Plain portion length not less than 0.5d. Includes the runout threads at the nut end
Nut end lengths   2d + 6mm for nominal lengths up to 125mm
2d + 12mm from over 125mm to 200mm nominal length
2d + 25mm over 200mm nominal length
For standard nominal lengths see BS 4439
Tolerance class for metal end thread to be 4h (close fit) or oversize. See BS 4439
Tolerance class for nut end thread to be 6g (medium fit)

STUDS

difference between a bolt and a screw is that a screw is threaded for virtually its entire length. The first thin thread is removed by a 45° chamfer extending to the minor diameter, or by rounding the end to a radius of approximately 1.25d. If the threads are rolled and not cut, the rolling operation forms a lead at the end of the thread and no other machining to remove the first thin thread is necessary.

Hexagon bolts, screws and nuts are identified as ISO metric by the symbol M or ISOM on the top face of the head of forged bolts and screws, or by the symbol M indented or rolled continuously along one face of the hexagon bar stock for turned nuts and bolts. In addition a symbol indicating the strength grade of the steel may appear. Full particulars are given in BS 3692.

BS 4439 covers metric studs for general purposes. A stud, shown on page 113, is a length of circular section bar, threaded at each end with a plain portion between. One end, called the metal end or fast end, is screwed into one detail of the assembly. The other end, called the nut end, passes through a plain hole in the second detail and carries the nut which locks the assembly together. The metal end threads are made either oversize or to tolerance class 4h (close fit), so that they are tight in the tapping. The nut end threads are made to tolerance class 6g (medium fit). The use of different tolerance classes ensures that when the nut is removed, the stud is not unscrewed as well. The length of the metal end thread, which includes the run-out threads, is d or 1.5d, where d is the thread diameter. The shorter length is used for steel, the longer for soft or brittle materials such as aluminium alloy or cast iron. The nut end length is calculated from the same formulae as those used to determine the thread lengths of hexagon bolts, and depends on the nominal length of the stud. This is the projecting length, that is, the sum of the plain portion and nut end lengths. BS 4439 lays down a series of standard nominal lengths from 12 mm to 500 mm.

Studs are used in situations where there is insufficient space on one side of the assembly for bolt heads or nuts. They are also used to secure details such as inspection covers which have to be removed frequently. The use of a nut and bolt in such an application would require two spanners to make the joint. If a set bolt were used, see page 115, the frequent removal and replacement would quickly wear the thread in the tapping, particularly if the material were cast iron or light alloy.

Nut and bolt assembly

A sectional view of such an assembly is given on page 115. Normally the details have clearance holes to facilitate assembly, although fitted bolts are also used. For drawing purposes clearance holes can be made 2 mm larger than the bolt diameter. BS 4186 gives clearance hole sizes for metric bolts and screws in close, medium and free fit series. The bolt

SCREW FASTENERS

Plain washer
Drg proportions
2d diameter
0.2d thick
See BS 4320

Clearance holes drawing size d+2mm dia
See BS 4186

a

f

a

d

b

c

Note
a. Clearance hole in each
   component
b,c. Thread each side of nut

NUT AND BOLT ASSEMBLY

e

a

b   c   d

Note
a. 120° drill point
b. Unthreaded bottom of
   tapping hole
c. Thread in tapping hole
   below bolt
d. Bolt thread above joint line
e. Clearance hole in outer
   detail

SET BOLT ASSEMBLY

d   e

a

b   c   f   g

Note
a. 120° drill point
b. Unthreaded bottom of
   tapping hole
c. Thread in tapping hole
   below stud
d. Metal end thread flush
   with joint line
e. Clearance hole in outer
   component
f,g Nut end threads above
   and below nut

NUT AND STUD ASSEMBLY

115

should be long enough to project right through the nut, so that the nut threads are fully engaged and all are sharing the tensile load in the bolt. Note that the nut and bolt are not sectioned although the section plane passes through them. If the details being joined are of soft material a plain washer may be placed under the nut, and sometimes under the bolt head as well, to prevent damage when the nut is tightened. Suitable drawing proportions for plain washers are given on page 115. For complete particulars BS 4320 should be consulted.

Set bolt assembly

A set bolt is sometimes called a tap bolt. It secures two details together by being screwed into a tapping in the inner. The outer detail has a clearance hole. To clamp the details together the underside of the bolt head must bear on the outer detail. To ensure this, the end of the bolt must not reach the bottom of the thread in the tapping, and some bolt threads must be above the joint line of the details. The correct representation of a blind tapped hole should also be noted. The tapping drill, whose diameter is drawn equal to the minor diameter of the thread, produces a hole with a conical end. The included angle of the cone varies with the material being drilled, but is usually 118°. For drawing purposes this is made 120°. The minor diameter drilling is threaded with a tap. It is difficult and relatively expensive to tap the hole right to the bottom, and generally a short length is left unthreaded. This accommodates the run-out threads produced by the lead-in taper on the tap, and must be shown on the drawing. These points are noted on page 115.

Nut and stud assembly

This is illustrated on page 113. A blind tapped hole is shown but through tappings are used in thin flanges. The stud is screwed into the tapping until the run-out threads at the metal end jam in the first thread in hole. To ensure that this occurs, the tapping depth in the hole is a few threads longer than the metal end length of the stud. Studs never bottom in the tapping. The outer detail has a clearance hole as with a set bolt, and the nominal length of the stud should be such that it stands through the nut, as with a nut and bolt assembly.

The forms and drawing proportions for other common screw fasteners are shown on pages 117 and 118. They are all known as screws, whether they are threaded up to the head or not. Machine screws may be used in tappings or with machine screw nuts, details of which are given in BS 4183. They are of the pressed type, and are made in square and hexagon forms.

Screws with hexagon socket heads are tightened by a special key or wrench made from hexagonal bar bent through 90° and called an Allen key. Cap screws with hexagon socket heads are often called Allen screws and are used extensively on machine tools and jigs and fixtures.

SCREW FASTENERS

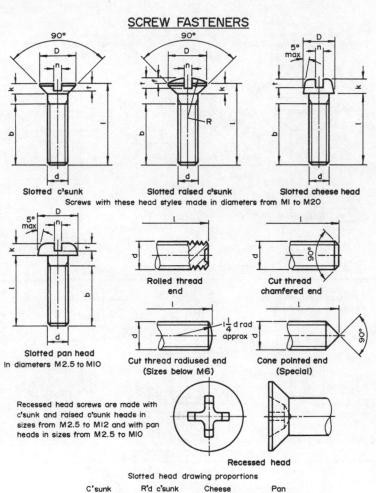

Slotted c'sunk	Slotted raised c'sunk		Slotted cheese head

Screws with these head styles made in diameters from MI to M20

Slotted pan head
In diameters M2.5 to MIO

Rolled thread end

Cut thread chamfered end

Cut thread radiused end
(Sizes below M6)

Cone pointed end
(Special)

Recessed head screws are made with c'sunk and raised c'sunk heads in sizes from M2.5 to MI2 and with pan heads in sizes from M2.5 to MIO

Recessed head

Slotted head drawing proportions

C'sunk	R'd c'sunk	Cheese	Pan
D= 2d	D= 2d = R	D=1.6d	D = 2d
k = 0.5d	k = 0.5d	k=0.6d	k = 0.6d
t = 0.3d	f = 0.25d	t = 0.4d	t = 0.6k
n = 0.25d	t = 0.5d	n = 0.25d	n = 0.25d
	n = 0.25d		

For standard nominal lengths l, thread lengths b and other particulars see BS 4183

MACHINE SCREWS

117

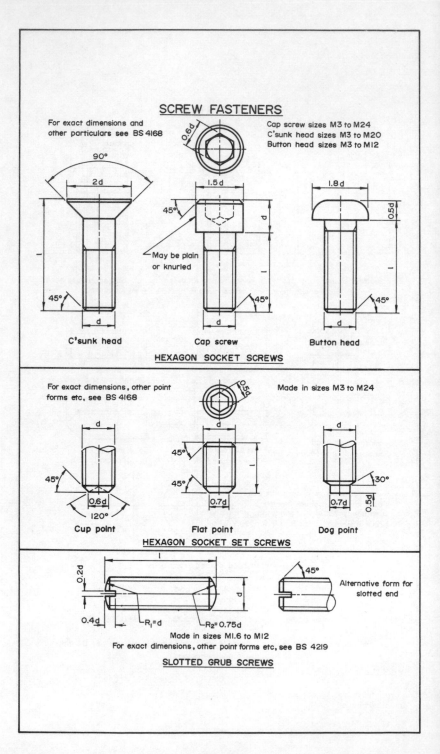

SCREW FASTENERS

For exact dimensions and
other particulars see BS 4168

Cap screw sizes M3 to M24
C'sunk head sizes M3 to M20
Button head sizes M3 to M12

90°

2d

1.5d

1.8d

0.6d

0.5d

45°

d

45°

May be plain
or knurled

l

l

l

45°

d

45°

d

45°

d

C'sunk head

Cap screw

Button head

HEXAGON SOCKET SCREWS

For exact dimensions, other point
forms etc, see BS 4168

Made in sizes M3 to M24

0.5d

d

d

d

45°

45°

l

30°

45°

0.5d

0.6d

0.7d

0.7d

120°

Cup point

Flat point

Dog point

HEXAGON SOCKET SET SCREWS

l

0.2d

45°

Alternative form for
slotted end

d

0.4d

R₁ = d

R₂ = 0.75d

Made in sizes M1.6 to M12
For exact dimensions, other point forms etc, see BS 4219

SLOTTED GRUB SCREWS

118

Hexagon socket set screws and slotted grub screws are used as locking devices which are screwed through one detail and have their ends bearing on, or inserted in, a second detail. Thus relative rotational movement between the two parts is prevented. A common application is for locking a door handle on to the lock spindle, when the dog type end fits into a hole drilled into the lock spindle. In cases where the screw end would damage the detail if allowed to bear on it directly, a small circular brass pad is used between the screw and the detail.

SCREW FASTENER PROBLEMS

Use First or Third Angle projection as required by the question. Draw full size unless otherwise stated. Do not show hidden detail.

1 Draw twice full size in First Angle projection, three views of an M20 hexagon nut as shown on page 112.

2 Using Third Angle projection draw three views of an M24 hexagon bolt with a washer faced head and a nominal length of 120 mm. Calculate the thread length from the formulae on page 115. Show the thread end chamfered.

3 Two pieces of material each 50 mm thick are to be secured together by an M16 hexagon bolt and nut. Draw a sectional view of the assembly similar to that shown on page 115. The nominal length of the bolt is to be 120 mm, the thread length is to be calculated, and a plain washer is to be used under the nut.

4 Draw a sectional view of a set bolt assembly as shown on page 115, with the following dimensions: Tapped detail 50 mm thick, outer detail 32 mm thick. Tapping M16, 28 mm minimum length of full thread. Other dimensions are to be settled by the student. The set bolt is to have a full bearing head.

5 Two pieces of light alloy, one 20 mm thick and the other 15 mm thick, are to be joined by an M6 stud and hexagon nut in a blind tapped hole. The nut is to be completely housed in a counterbore. Determine suitable dimensions for the assembly and draw twice full size a sectional view similar to that on page 115. Add an end view looking on the nut, using Third Angle projection. Show a plain washer under the nut.

6 An incomplete plan and elevation of a clamping arrangement are shown in Figure 1 on page 121. The clamp is secured to the base plate by an M12 stud and hexagon nut. The elliptical flange, which is to be

drawn by an approximate method, is located on the 50 mm diameter spigot on the base plate. Using Third Angle projection, draw a sectional elevation of the arrangement on AA, and a complete plan.

7 Figure 2 shows a blanking cover for a 60 mm diameter hole in a casting. The cover is to be secured by four M16 set bolts with washer faced heads and spring washers. Draw in First Angle projection a half sectional front elevation corresponding to the given views, with the section plane passing through one bolt, and a half plan.

8 Views of a small pipe support are given in Figure 3. The cap is secured by an M6 stud and hexagon nut on one side, and by an M6 hexagon bolt and nut on the other. Draw a sectional elevation on AA and a plan of the complete assembly in First Angle projection.

SCREW FASTENER PROBLEMS

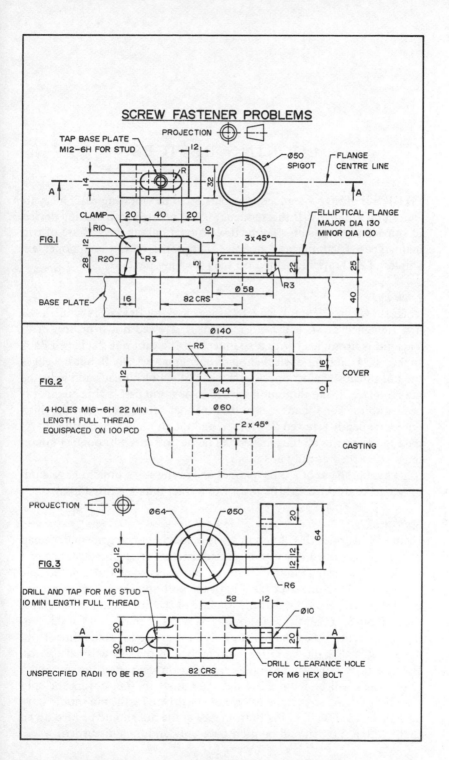

PROJECTION

TAP BASE PLATE M12-6H FOR STUD

Ø50 SPIGOT

FLANGE CENTRE LINE

CLAMP

12

14

32

A

A

20   40   20

ELLIPTICAL FLANGE MAJOR DIA 130 MINOR DIA 100

FIG.1

R10

12

28   R20   R3

10

3 x 45°

5

22

25

R3

Ø58

40

BASE PLATE

16

82 CRS

Ø140

R5

16

12

FIG.2

Ø44

10

COVER

Ø60

4 HOLES M16-6H 22 MIN LENGTH FULL THREAD EQUISPACED ON 100 PCD

2 x 45°

CASTING

PROJECTION

FIG.3

Ø64   Ø50

20   64

20   12 12

12

R6

DRILL AND TAP FOR M6 STUD 10 MIN LENGTH FULL THREAD

58   12

Ø10

A   20   A

20   R10

20

DRILL CLEARANCE HOLE FOR M6 HEX BOLT

UNSPECIFIED RADII TO BE R5

82 CRS

121

11

LOCKING DEVICES

THERE is always a tendency for nuts fitted to vibrating machinery to slacken off gradually. If this tendency is not resisted by a locking device of some kind the results may be disastrous. Locking devices are of two main types, frictional and positive, and some of the commoner examples of each are illustrated on the following pages.

Lock nut

This, shown in Figure 1, is a frictional locking device. A thin nut is first tightened down and a normal nut is screwed down on top of it. This nut is firmly held with a spanner and the thin nut slackened back slightly. The amount of slackening is of course very small, but it wedges the nut threads against opposite flanks of the bolt threads and jams the nuts in place. If the thin nut is placed in position first, a thin spanner is needed to slacken it back. Since thin spanners are uncommon the thick nut is frequently screwed down first, and the thin nut placed on top. As the top nut carries all the tensile load in the bolt this, although common practice, is theoretically unsound.

To avoid the use of a thin spanner and at the same time to have a nut of adequate thickness at the top, two normal nuts are often used.

Split pin

Figure 2 illustrates this common positive locking arrangement. The end of the bolt is turned down to a plain diameter for a short distance and a hole drilled through it as shown. A split pin is driven through the hole and the legs bent one over the end of the bolt and the other over a flat on the nut. It is essential that the top face of the nut should bear against the split pin and that the pin should be a good fit in the hole or slackening may occur. When the bolt has been in service for some time it will have stretched slightly and the bottom face of the nut will have worn. Because of this the split pin may not bear on the top of the nut and a thin washer will be required under the nut. If the nut covers the split pin hole partially so that the pin cannot be driven in, the pin should have a small flat filed on it, or the bottom face of the nut should be filed down slightly. Split pins should be used once only and then discarded.

Lock washers

Two forms of these very common frictional locking devices are shown in Figure 3. Spring washers are wound from square or rectangular section wire and the single coil type is illustrated. A double coil rectangular section type is also made which is useful as an anti-rattle device for applications involving light alloys.

When a nut is tightened down on a spring washer the washer thrusts the nut threads against the bolt threads, thus increasing the friction between them. One type has deflected ends at the split, producing sharp corners which dig into the nut and detail and increase the locking effect. This type damages the detail if used alone and a thin washer is generally used under them. Suitable drawing proportions for spring washers are outside diameter 2d and thickness 0·2d, where d is the major diameter of the thread.

A variation of the spring washer is the star washer. Internal or external projections are twisted to form many sharp corners which dig into the detail and the bottom of the nut.

Taper pin

Figure 4 shows a nut locked on to a bolt by a taper pin which passes right through them. This is a very safe positive locking device particularly if the small end of the taper pin is split and the legs are opened out as shown. A disadvantage is that if the nut is removed it is very difficult to line up the holes in the nut and bolt again. The hole also weakens both details.

Slotted nut

This is a hexagon nut with six slots cut across the top face as shown in Figure 5. A split pin passes through one slot and a hole in the bolt, the legs of the pin being opened and bent round the faces of the nut. This gives a positive lock. It is sometimes difficult to line up a slot with the hole in the bolt, and the hole weakens the bolt. However, the height of the nut is increased compared with that of a normal nut, so that the slots do not weaken it.

Castle nut

Figure 6 shows this development of the slotted nut. The top of the nut carries a cylindrical rim into which six or eight slots are cut, depending on the size of the thread. A split pin passes through one slot and the hole in the bolt, and the legs of the pin are bent round the rim. With this nut the bent legs of the split pin are safer than with a slotted nut as they do not stand out on the nut faces.

LOCKING DEVICES

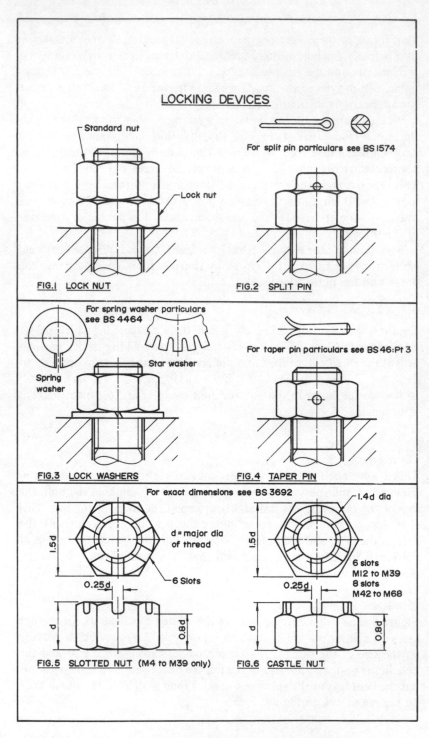

For split pin particulars see BS 1574

FIG.1   LOCK NUT

Standard nut

Lock nut

FIG.2   SPLIT PIN

For spring washer particulars see BS 4464

Star washer

Spring washer

FIG.3   LOCK WASHERS

For taper pin particulars see BS 46:Pt 3

FIG.4   TAPER PIN

For exact dimensions see BS 3692

1.5d

d = major dia of thread

6 Slots

0.25d

d

0.8d

FIG.5   SLOTTED NUT (M4 to M39 only)

1.4d dia

1.5d

6 slots M12 to M39
8 slots M42 to M68

0.25d

d

0.8d

FIG.6   CASTLE NUT

LOCKING DEVICES

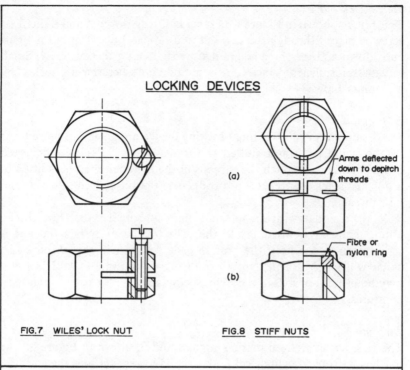

(a)

Arms deflected down to depitch threads

(b)

Fibre or nylon ring

FIG.7 WILES' LOCK NUT

FIG.8 STIFF NUTS

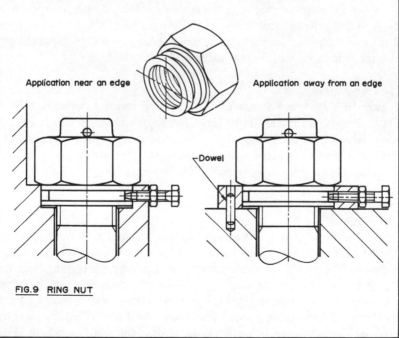

Application near an edge

Application away from an edge

Dowel

FIG.9 RING NUT

Wiles' lock nut

This nut, as shown in Figure 7, is sawn half way through and a small set screw is placed through the saw cut in a tapped hole. The nut is tightened down and the threads jammed by tightening the set screw. Small nuts with insufficient space to accommodate a set screw are locked by a hammer blow.

Stiff nut

Two examples of this frictional locking device are given in Figure 8. At (a) the cylindrical rim is slotted to form two arms which are deflected down slightly to de-pitch the threads in them. When the nut is fitted to a bolt the arms are forced up, producing friction between the nut and bolt threads.

At (b) a collar with an internal diameter smaller than the thread minor diameter is fitted inside the cylindrical rim. When the nut is screwed on to the bolt, the bolt threads force their way through the collar which grips them tightly producing a frictional lock. The collar may be of fibre or nylon, the nylon being suitable for use at high temperatures.

Ring nut

The ring nut is used on marine engines and is shown in Figure 9. The nut has a grooved cylindrical collar on its lower face and the end of a set screw fits into the groove. The set screw is locked by a hexagon nut. Near an edge the collar is housed in a counterbore. Away from an edge it fits in a ring which is prevented from rotating by a dowel. A split pin is often fitted as an additional safeguard.

Locking plates

Figure 10 shows two forms of this positive locking device, which is made from sheet metal. They have bi-hexagonal holes which engage with the hexagon on the nut. The plate is secured to the main component by a small set screw, which often has its own locking device. Locking plates may have a half hole, or they may enclose the nut completely.

Tab washers

Three common forms of this positive locking device are shown in Figures 11 and 12. Figure 11(a) shows a type suitable for fitting over the edge of a flange. The washer is fitted under the nut, one leg is bent over the flange and the other up against a flat on the nut. If no flange is available, the form shown in Figure 11(b) can be used. Here a plug is shown locked by a tab washer having a small leg fitting in a hole drilled in the component, the other leg being bent up against the plug as before. This

126

LOCKING DEVICES

A spring washer or other locking device may be provided for the set screw

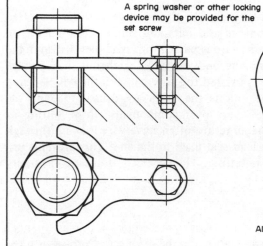

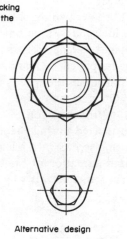

Alternative design

FIG.10 LOCKING PLATE

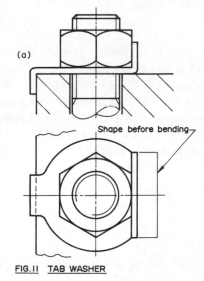

(a)

Shape before bending

FIG.11 TAB WASHER

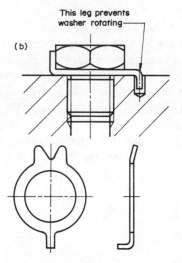

(b)

This leg prevents washer rotating

second leg has a vee cut in it, so that it can be bent against two faces of the hexagon if the final position of the plug requires it. Figure 12 shows a Siamese tab washer which is two tab washers made in one piece. Legs to prevent rotation of the washer are unnecessary. Siamese tab washers can be used only on faces with no projections between the bolt holes.

Wire locking

Figure 13 shows two set bolts locked by this positive method. It is often used to lock plugs, pipe unions and nuts and similar components on aero-engines. A length of soft wire is passed through holes drilled in the bolt heads, pulled taut and the ends twisted together with a pair of pliers. Sometimes the wire is twisted together between the details being locked. On nuts and pipe unions the holes are drilled through the corners of the hexagon. Covers over components whose setting or position is fixed during manufacture often have wire locking through the securing screws with a lead seal fixed to the end. This prevents unauthorized alteration of the setting. The covers of domestic electricity meters are sealed in this way.

Grub screw

The circular nut, shown in Figure 14, is tightened by a C or claw spanner and locked by a grub screw. Tightening the grub screw presses a brass pad against the bolt threads giving a frictional lock. The pad prevents the grub screw end damaging the bolt threads.

Serrated locking plate

Many locking devices have been designed for special applications, and the locking plate shown in Figure 15 is an example. The plate is free to pivot about its centre, and is spring-loaded to keep it against the face of the component. The serrations in the plate engage with serrations around the base of the nuts. The centre line of the plate is at a small angle to the centre line of the nuts, so any tendency for one nut to unscrew tends to tighten the other. The plate is engaged or disengaged by lifting it against the spring with a hook through one of the lifting holes. It can then be turned about its centre.

LOCKING DEVICES

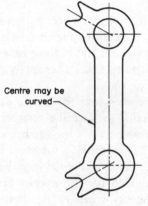

Centre may be curved

FIG.12 SIAMESE TAB WASHER

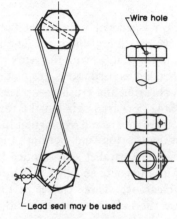

Direction of wire run prevents slackening of bolts

Wire hole

Lead seal may be used

FIG.13 WIRE LOCKING

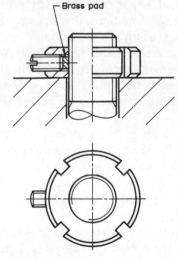

Brass pad

FIG.14 GRUB SCREW

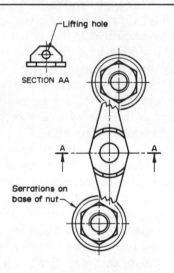

Lifting hole

SECTION AA

Serrations on base of nut

FIG.15 SERRATED LOCKING PLATE

12

KEYS AND COTTERED JOINTS

A KEY is a detail inserted axially between a shaft and a hub to prevent relative rotation of the parts. The hub may slide along the shaft whilst rotating with it, as with the gears in a car gearbox, and in these cases a feather key is used. Keys are made from steel to resist adequately the high shearing and crushing stresses placed on them.

BS 4235: Part 1 sets out the forms and dimensions for metric keys which are square and rectangular in cross-section with parallel or tapered faces at the top and bottom. Figure 1 opposite shows square and rectangular parallel keys. Notice that the key is sunk in the shaft for half its thickness and that it is fitted to the sides of the keyways with top clearance. Three classes of fit are provided for these keys: free, where the hub is required to slide over the key when in use, that is, where the key is used as a feather; normal, where the key is to be inserted in the keyway with the minimum fitting, as in mass production assembly; close, where an accurate fit of key is required. Here fitting will be necessary under maximum material conditions, that is, when the keyway width is a minimum and the key width is a maximum.

Keys are not normally supplied with the ends radiused as in Forms A and C, or with chamfers. The chamfers are necessary to prevent the corners of the key fouling the radii in the bottom of the keyway. Sharp corners in the keyway would act as stress raisers with a consequent risk of fracture at these points.

Form A keys are used in keyways machined with an end mill, the keyway being situated part way along the shaft. The semicircular ends of the key fit the corresponding ends of the keyway produced by this type of cutter. Form C keys are used in similar keyways cut at the end of a shaft, which have only one semicircular end. Form B keys for are use in keyways cut with a slotting saw.

Parallel keys are used to transmit unidirectional torques where heavy starting loads are not involved, and where periodic withdrawal or sliding of the hub may be required. When a gib head, see below, cannot be accommodated and there is insufficient space to drift out the key from behind, the hub must be withdrawn over the key and a parallel key is essential. Reference to the table on page 223 will show that square keys are used for shafts up to 22 mm diameter. Over this size rectangular

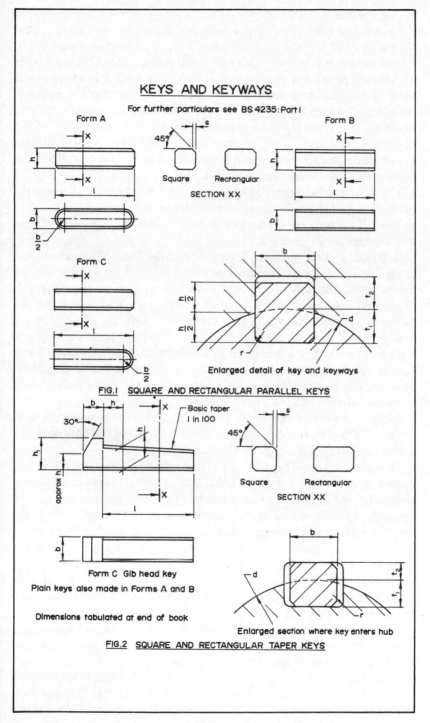

KEYS AND KEYWAYS

For further particulars see BS 4235:Part I

Form A

45°

Square   Rectangular

SECTION XX

Form B

h

b

$\frac{b}{2}$

Form C

$\frac{b}{2}$

b

$\frac{h}{2}$

$\frac{h}{2}$

t_2

d

t_1

r

Enlarged detail of key and keyways

FIG.I   SQUARE AND RECTANGULAR PARALLEL KEYS

30°

b   h

Basic taper
I in IOO

h_1

approx h

45°

Square   Rectangular

SECTION XX

l

b

Form C Gib head key

Plain keys also made in Forms A and B

Dimensions tabulated at end of book

b

d

t_2

t_1

r

Enlarged section where key enters hub

FIG.2   SQUARE AND RECTANGULAR TAPER KEYS

131

keys are used.

Square and rectangular taper keys are illustrated in Figure 2, and have a taper of 1:100 on the top face. They are made as plain keys in Forms A and B and with a gib head in Form C. The gib head makes withdrawal of the key possible by driving a tapered drift between the back of the head and the hub. Taper keys fit at the top and bottom, with side clearance. They are used to transmit heavy unidirectional, reversing and vibrating torques, and where periodic withdrawal of the key may be necessary. They cannot be used as feathers. As with parallel keys, square taper keys are used for shafts up to 22 mm diameter, with rectangular keys being used over this size.

The relations between shaft diameter and key section given in the table on page 223 are for general applications. Smaller key sections may be used if suitable for the torque to be transmitted. The use of larger key sections is not permitted.

A Woodruff key, shown in Figure 3, is in the form of a segment of a circle and fits in a corresponding recess in the shaft. It adjusts itself to any taper on the hub keyway and is widely used on machine tools. It cannot be used as a feather because it may jam, and it has the disadvantage that the deep keyway weakens the shaft.

Figure 4 shows two types of round key. The plain type is a length of circular bar which is a driving fit in a hole drilled half in the shaft and half in the hub. Round keys may also be threaded and screwed into position, in which case they usually have a square head for fitting which is machined off when the key is in place.

Saddle keys are of two types, flat and hollow, and are illustrated in Figure 5. They are suitable only for light torques, and the hollow saddle key is used solely for temporary fastenings.

Splines, shown in Figure 6, are projections of rectangular cross-section machined on a shaft. In effect they are several keys integral with the shaft, and fit corresponding recesses in the hub. They are equally spaced round the shaft and vary in number from four upwards. If the angular relationship between shaft and hub is important, one spline may be left uncut on the shaft and one removed from the hub. The shaft and hub will then assemble in one position only. Serrations are similar to splines but are triangular in cross-section instead of rectangular. Since they are smaller than splines the angular relationship between shaft and hub may be closely controlled. For this reason they are often used to secure levers to shafts, when the lever position is to be altered for adjustment purposes. Splined and serrated shafts are used extensively in the car and aeroplane industries.

The recommended ways of dimensioning keyways are shown in Figure 7. Note that the drawings are not completely dimensioned. Only those dimensions are given for which alternatives might be chosen.

KEYS AND KEYWAYS

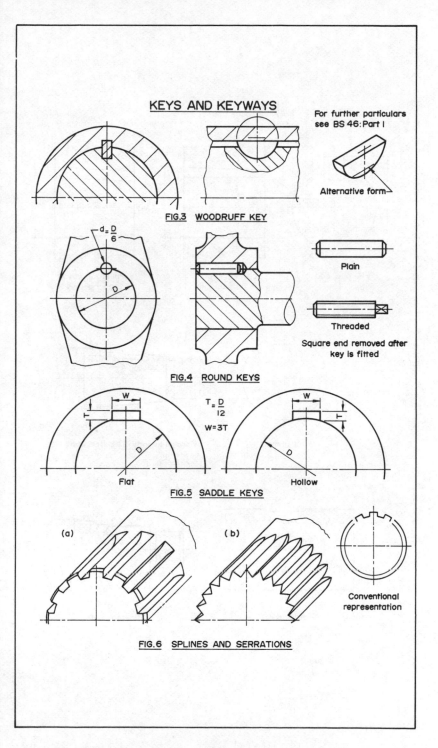

For further particulars
see BS 46:Part I

Alternative form

FIG.3   WOODRUFF KEY

$d = \dfrac{D}{6}$

Plain

Threaded

Square end removed after
key is fitted

FIG.4   ROUND KEYS

$T = \dfrac{D}{12}$

$W = 3T$

Flat   Hollow

FIG.5   SADDLE KEYS

(a)   (b)

Conventional
representation

FIG.6   SPLINES AND SERRATIONS

133

KEYS AND KEYWAYS

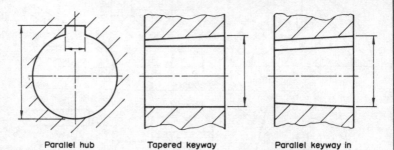

Parallel hub   Tapered keyway   Parallel keyway in
                        in parallel hub   tapered hub

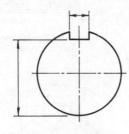

Parallel shaft   Parallel keyway in tapered shaft

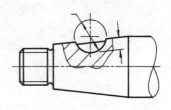

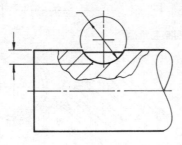

Parallel shaft   Tapered shaft

FIG.7   DIMENSIONING OF KEYWAYS

134

Cottered joints

These are used to join rods which carry axial loads only and Figure 1 on page 136 shows a typical arrangement. One rod end fits in a socket in the other and both rods are slotted to take the cotter. This is a flat piece of material with a taper on one side. The slots are a little longer than the width of the cotter, and their positions are such that driving the cotter in pulls one rod into the socket in the other. This is shown by the arrows on the elevation. The clearances on each side of the cotter are important and should be noted. Figure 1 shows the rod end and socket tapered, but they may be cylindrical. Also the cotter can have square instead of round ends. The ends of the cotter which take the hammer blows are chamfered to reduce spreading and cracking. Cottered joints are easily assembled and dismantled and the details always take up the same positions when the joint is remade.

Other applications of cottered joints are shown in Figures 2, 3 and 4. Figure 2 shows a cotter pin securing a lever to a spindle. This method is used to secure a bicycle crank to the chain wheel spindle. A rod secured to a plate by a cotter is illustrated in Figure 3. Figure 4 shows a strapped joint secured by a gib and cotter. The gib prevents the lower end of the strap spreading as the cotter is driven in, and allows a parallel hole to be used in the rod end and strap.

COTTERED JOINTS

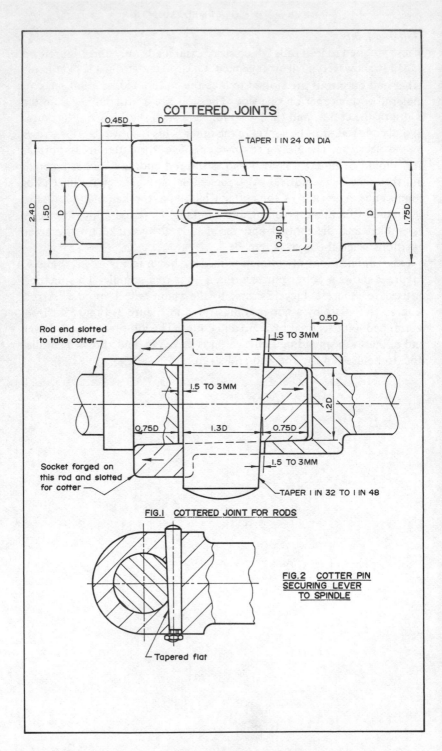

TAPER 1 IN 24 ON DIA

0.45D

D

2.4D

1.5D

D

0.3ID

1.75D

Rod end slotted to take cotter

1.5 TO 3MM

0.5D

1.5 TO 3MM

1.2D

0.75D

1.3D

0.75D

Socket forged on this rod and slotted for cotter

1.5 TO 3MM

TAPER 1 IN 32 TO 1 IN 48

FIG.1   COTTERED JOINT FOR RODS

FIG.2   COTTER PIN SECURING LEVER TO SPINDLE

Tapered flat

COTTERED JOINTS

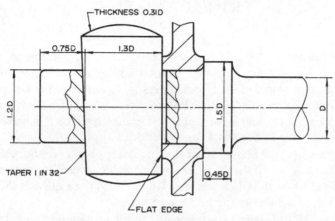

THICKNESS 0.31D

0.75D 1.3D

1.2D

1.5D

D

TAPER I IN 32

0.45D

FLAT EDGE

FIG.3   COTTERED JOINT FOR ROD AND PLATE

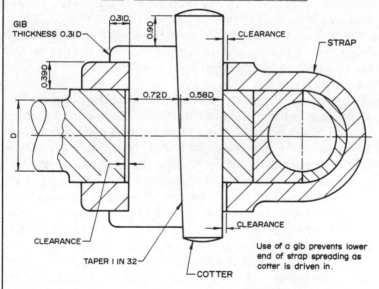

GIB
THICKNESS 0.31 D

0.31D

0.9D

CLEARANCE

STRAP

0.39D

0.72D   0.58D

D

CLEARANCE

TAPER I IN 32

COTTER

CLEARANCE

Use of a gib prevents lower
end of strap spreading as
cotter is driven in.

FIG.4   GIB AND COTTER JOINT FOR STRAP

137

13

TRUE LENGTHS

IT is important to be able to find the true length of a line in some aspects of Engineering Drawing, particularly in the development, or laying out in a plane, of sheet metal details. It was stated in Chapter 5 on orthographic projection that the projection of a line will only show its true length if the line is parallel to the plane on which it is projected. This is illustrated in Figures 1 (a), (b) and (c) opposite.

In Figure 1 (a) the plan $a^2 b^2$ of the line AB is parallel to the XY line. Hence the line is parallel to the vertical plane of projection, and its elevation $a^1 b^1$ on this plane is a true length. Also the angle θ is the true inclination of the line to the horizontal plane.

The line AB in Figure 1 (b) is parallel to the horizontal plane. In this case, therefore, the plan $a^2 b^2$ of the line shows its true length, and the angle ϕ its true inclination to the vertical plane.

Figure 1(c) shows the elevation $a^1 b^1$ of the line AB to be parallel to the auxiliary vertical plane, and so the projection of the line on this plane gives its true length. Angle θ is the true inclination of the line to the horizontal plane.

When a line is not parallel to one of the normal planes of projection two methods may be used to find its true length. The line may be moved until it is parallel to a principal plane, or an auxiliary plane may be used which is parallel to the plan or elevation of the line. Figure 2 illustrates the first of these approaches, which is called the revolution method.

In both illustrations the end A of the line remains stationary and the end B is revolved until the line is parallel to either the horizontal or vertical plane. Thus the line sweeps out a right cone with its apex at A. In one view B moves in a circle; in the other it moves along a straight line. There are two positions for the true length, depending on the direction of rotation of B. If the true length appears in the elevation, then the angle it makes with the XY line is the true inclination of the line to the horizontal plane. If it appears in the plan, its angle to XY is the true inclination to the vertical plane, since, relative to the plan, the XY line represents the vertical plane, and vice versa.

Finding the true length of a line by triangulation, as in Figure 3, is essentially the same as the revolution method, but for some purposes it is more convenient. The plan length $a^2 b^2$ of the line is set off at right

TRUE LENGTHS

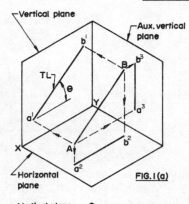

FIG.1(a)

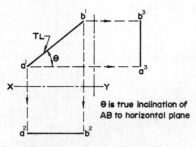

Θ is true inclination of AB to horizontal plane

Line AB is parallel to vertical plane so projection on that plane is true length.

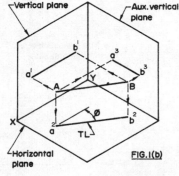

FIG.1(b)

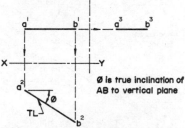

Ø is true inclination of AB to vertical plane

Line AB is parallel to horizontal plane so projection on that plane is true length

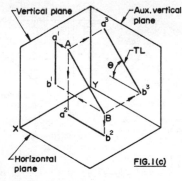

FIG.1(c)

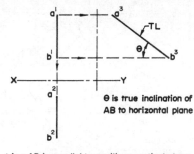

Θ is true inclination of AB to horizontal plane

Line AB is parallel to auxiliary vertical plane so projection on that plane is true length

139

TRUE LENGTHS

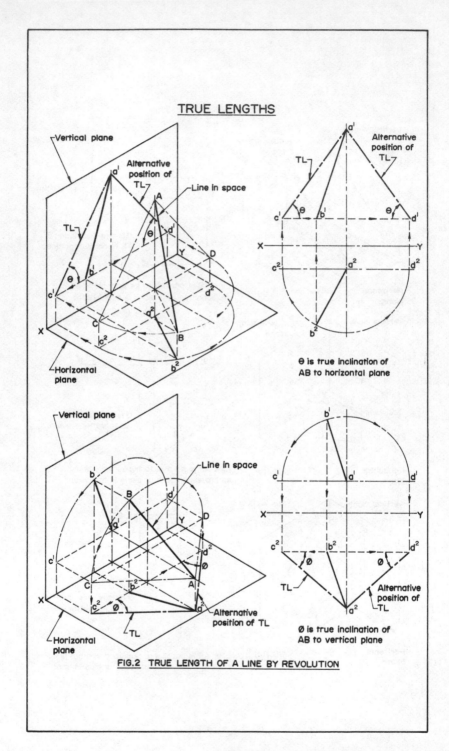

Θ is true inclination of AB to horizontal plane

Ø is true inclination of AB to vertical plane

FIG.2   TRUE LENGTH OF A LINE BY REVOLUTION

angles to the difference in vertical height of the ends of the line in elevation. The hypotenuse AB of the right angled triangle so formed is the true length of the line, and angle θ is the true inclination to the horizontal plane. The true length will also be found as the hypotenuse of a right angled triangle formed by setting off the elevation $a^1 b^1$ at right angles to the difference in depth of the ends of the line in plan. In this triangle the true inclination of the line to the vertical plane will appear.

The second approach, using auxiliary planes, is shown in Figure 4. The plan $a^2 b^2$ of the line is viewed normally in the direction of arrow S, and a view in this direction is drawn on an auxiliary vertical plane $X^1 Y^1$. This plane is parallel to $a^2 b^2$, so the auxiliary elevation drawn on it will show the true length of AB. The heights h and h^1 are the same in both elevations, and θ is the true inclination of the line to the horizontal plane.

If the elevation $a^1 b^1$ of the line is viewed normally in the direction of arrow T, an auxiliary plan projected on $X^1 Y^1$, which is parallel to $a^1 b^1$, will again show the true length of the line. The widths w and w^1 from the original plan are the same in the auxiliary plan. ϕ is the true inclination of the line to the vertical plane. For a fuller treatment of the principles of projecting auxiliary views the student is referred to Book 2.

In the worked examples which follow the methods for finding the true length of a line have been used to draw orthographic views of lines and flat laminae in various positions relative to the planes of projection.

Example 1
Sufficient information is given to draw three orthographic views of the lines. The true lengths of AB and BC have been found by the revolution method and that of AC by triangulation. Since the three true lengths all appear in elevations, the angles they make with the XY line are their true inclinations to the horizontal plane.

Example 2
AB can be positioned in the three views directly from the given data. Its front elevation $a^1 b^1$ is true length. To find c^1 imagine the triangle to be rotated about $a^1 b^1$ until it is parallel to the vertical plane. In this position it can be drawn, since it will show its true size and shape $a^1 b^1 c^4$. When the triangle is rotated about $a^1 b^1$, c^4 will appear to move along a line perpendicular to $a^1 b^1$. Where this line meets the horizontal plane is the position of c^1. Project from c^1 into the plan and fix c^2 on the projector by striking an arc from a^2 equal to $a^1 c^4$, the true length of AC. This can be done because AC is a true length in plan. The drawing can now be completed by positioning c^3 in the end view.

TRUE LENGTHS

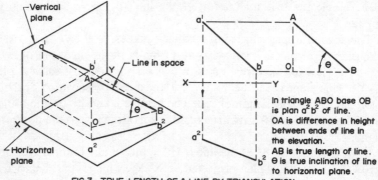

In triangle ABO base OB
is plan $a^2 b^2$ of line.
OA is difference in height
between ends of line in
the elevation.
AB is true length of line.
Θ is true inclination of line
to horizontal plane.

FIG.3 TRUE LENGTH OF A LINE BY TRIANGULATION

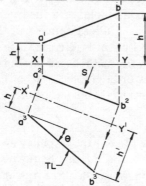

Auxiliary elevation is drawn by
viewing plan normally in direction
of arrow S.
Heights h and h^1 in auxiliary
elevation are equal to heights
in original elevation.
Θ is true inclination of line to
horizontal plane.

Auxiliary plan is drawn by
viewing elevation normally
in direction of arrow T.
Widths w and w^1 are the
same in both plan views.
$\emptyset$ is the true inclination of
line to vertical plane.

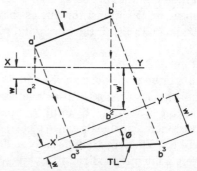

FIG.4 TRUE LENGTH OF A LINE BY AUXILIARY VIEWS

TRUE LENGTHS

Example 1

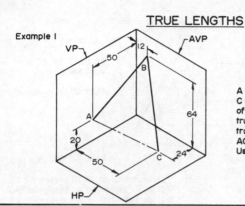

A is on the VP, B on the AVP and C on the HP. Find the true lengths of the lines AB and BC and the true distance AC. Also find the true inclinations of AB, BC and AC to the horizontal plane. Use First Angle projection.

Example 2

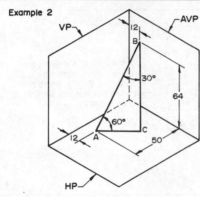

ABC is a right angled triangle positioned with AC on the horizontal plane and B touching the auxiliary vertical plane. Draw in First Angle projection an elevation, plan and end view of the triangle in this position.

Example 3

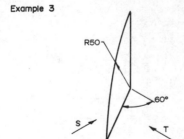

One edge of the quadrant is horizontal. Using Third Angle projection draw an elevation in the direction of arrow T, an end view in the direction of arrow S and a plan.

TRUE LENGTHS

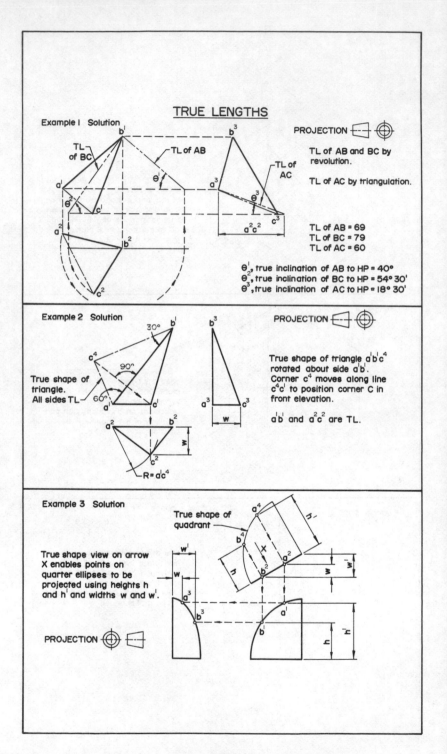

Example 1 Solution

TL of BC
TL of AB
TL of AC

θ^1
θ^2
θ^3

a^2c^2

PROJECTION

TL of AB and BC by revolution.

TL of AC by triangulation.

TL of AB = 69
TL of BC = 79
TL of AC = 60

θ^1, true inclination of AB to HP = 40°
θ^2, true inclination of BC to HP = 54° 30'
θ^3, true inclination of AC to HP = 18° 30'

Example 2 Solution

30°
90°
60°

True shape of triangle.
All sides TL.

R = a^1c^4

PROJECTION

True shape of triangle $a^1b^1c^4$ rotated about side a^1b^1. Corner c^4 moves along line c^4c^1 to position corner C in front elevation.

a^1b^1 and a^2c^2 are TL.

Example 3 Solution

True shape of quadrant

True shape view on arrow X enables points on quarter ellipses to be projected using heights h and h' and widths w and w'.

PROJECTION

144

Example 3

The plan view of the quadrant, which is a straight line, can be drawn from the given information. The elevation and end view project as quarter ellipses, and to draw them an auxiliary view in the direction of arrow X is required. This view shows the true shape of one quadrant. On the true shape select points such as a^4 and b^4, project them to the plan at a^2 and b^2, and then to the elevation at a^1 and b^1. The heights h and h^1 from the auxiliary view fix a^1 and b^1 on the projectors from the plan. Widths w and w^1 from the plan position a^3 and b^3 in the end view. The elevation and end view can then be lined in with a french curve.

TRUE LENGTH PROBLEMS

Draw full size using First or Third Angle projection as required by the question.

1 Two lines AB and BC are shown with A on the vertical plane, B on the auxiliary vertical plane and C on the horizontal plane. Draw three views in First Angle projection of the lines on these planes, and determine the true length of AB, AC and BC and the true angle θ.

2 The triangle is positioned with AC on the horizontal plane and B against the vertical plane. Draw three views in First Angle projection of the triangle on the given planes.

3 The given triangle has A on the vertical plane, B on the auxiliary vertical plane and C on the horizontal plane. Project views of the triangle on these planes, using First Angle projection.

4 The quadrant is positioned as shown with AB on the vertical plane, parallel to the horizontal plane. Draw views of the quadrant on the planes using First Angle projection.

5 The triangle has BC on the horizontal plane and A on the auxiliary vertical plane. Draw a front elevation, plan and end view in First Angle projection of the triangle in the given position.

6 Using First Angle projection draw three views of the semicircle. A is on the auxiliary vertical plane, B is on the horizontal plane, and the semicircle touches the vertical plane.

7 The plan view of a piece of plate is given, the plate being bent along the line BC. BC lies on the horizontal plane and D is 20 mm above the horizontal plane. The true sizes of angles B and C are 60° and 30° respectively. Use Third Angle projection to draw the given plan and an elevation looking in the direction of arrow R. Determine the

shape of the plate before bending and the true inclination of the edge AB to the horizontal plane.

8 A piece of sheet metal is bent as shown and has points D and B on the horizontal plane, A on the vertical plane and C on the auxiliary vertical plane. Draw three views of the plate on the given planes in First Angle projection, and construct its shape before bending.

9 Three orthographic views of a line are given in the figure. Draw these views and position a point C in each of them, such that the true lengths of AC and BC are 82 mm and 95 mm respectively, with C on the horizontal plane. Use Third Angle projection.

10 The plan and elevation of a triangular pyramid are given. Draw these views in Third Angle projection and find the true lengths of AB, AC and AD, the true inclinations of AB and AC to the horizontal plane, and the true inclination of AD to the vertical plane.

11 Using Third Angle projection reproduce the three given views of the triangle.

12 Views are given of a sheet metal component with an open top and bottom. Draw these views in Third Angle projection and construct the shape of half the component before it is bent.

TRUE LENGTH PROBLEMS

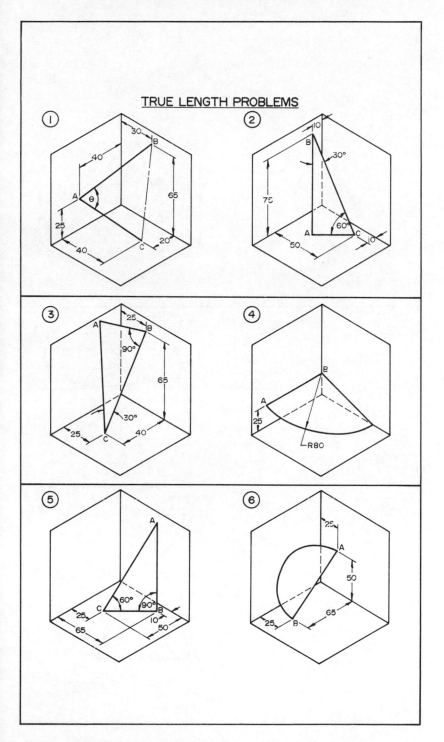

TRUE LENGTH PROBLEMS

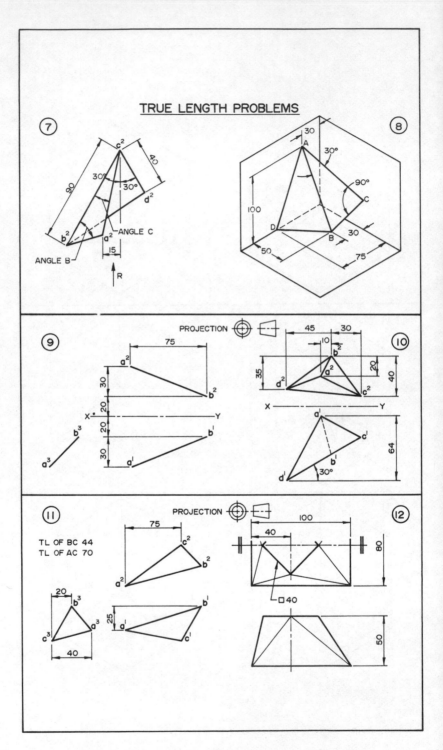

⑦

c²
90
40
30°
30°
d²
b²
a²
ANGLE C
ANGLE B
15
R

⑧

30
A
30°
90°
C
100
D
B
30
50
75

PROJECTION ⊕ ⟋◱◲

⑨
75
a²
30
b²
X
20
Y
20
b¹
b³
30
a¹
a³

⑩
45
30
10
b²
a²
20
35
d²
40
c²
X
a¹
Y
64
c¹
d¹
30°
b¹

PROJECTION ⊕ ⟋◱◲

⑪
TL OF BC 44
TL OF AC 70
75
c²
b²
a²
20
b³
c³
a³
40
b¹
25
a¹
c¹

⑫
100
40
80
□40
50

148

14

ISOMETRIC PROJECTION

ISOMETRIC projection is a method of producing a pictorial view of an object, that is, a view which shows three faces of the object simultaneously.

Figure 1(a) shows a cube positioned with its body diagonal horizontal, in front of a vertical plane. When the cube is in this position all its edges are at the same angle to the projection plane. If the cube is projected orthographically on to the plane, all the edges will project as the same length, that is, they will be equally foreshortened. From this fact comes the name of the projection system, iso – equal, metric – measure.

Figure 1(b) shows the projected view, and it is found that all the cube edges appear at 30° to the horizontal, except AD, OC and BF which appear vertical. Thus, edges which are parallel on the cube or other object being drawn, remain parallel in the isometric projection. This makes the view easy to draw, but since no account is taken of perspective it is slightly distorted. However, the distortion becomes objectionable only when very long objects are drawn.

The lines OA, OB and OC, redrawn in Figure 1(c), are called the isometric axes, and on them the drawing is built up. Alternative positions of the axes may be used, as will be seen later, these positions resulting from different orientations of the object to the projection plane. The 120° angles between the axes remain the same though, whatever position is used. Lines parallel to the axes are called isometric lines; others are non-isometric lines. The planes OADC, OAEB and OBFC are the isometric planes.

It should be noted that in preparing an isometric drawing, measurements may only be scaled directly on to isometric lines. This may be verified by considering the diagonals AB and EO in Figure 2(a). These lines, which are non-isometric lines, have the same true length, but their isometric lengths are obviously not equal. However, it should be noted that AB, which is parallel to the projection plane, is in fact a true length.

Isometric scale
Since all the edges of the cube are equally foreshortened, it follows that a view made by using the true lengths of the edges will be larger than

the object actually is. To obtain a full size view the isometric scale must be used.

Consider Figure 2(a) which again shows an isometric view of a cube. If the top face AEBO is rotated about AB until it is vertical, its true shape AE'BO' is obtained. In this view angle E'AO is 45° and AE' is the true length of the edge of the cube. The amount of foreshortening is the difference between this true length and the isometric length AE. This fact may be used to produce the isometric scale as shown in Figure 2(b). The ratio of isometric length to true length is $\sqrt{2}$ to $\sqrt{3}$ or 0.816 to 1.

In practice little use is made of the isometric scale. A drawing made using the natural scale will be correctly proportioned and this is more important than its being oversize.

Strictly, a view drawn using the isometric scale is an isometric projection, since this is what is produced by projecting the object orthographically as in Figure 1(a). A view drawn using the natural scale is correctly called an isometric drawing. However, the distinction is often ignored in practice and the two terms used indiscriminately.

Several worked examples follow which demonstrate how features such as angles, curves and circles are treated.

Objects composed entirely of isometric lines
These objects are easily drawn as all measurements in the orthographic views may be scaled directly on to the isometric view. It is unnecessary to draw the orthographic views. Example 1 is of this type. The solution illustrates the convenient technique of first drawing a box which will just contain the object, and then building up the shape of the object inside the box. Most objects can be treated in this way, and fewer errors are made by beginners who use it.

Objects with non-isometric lines
Lines on an object which are located by angles are non-isometric lines. Angles cannot be laid off directly on an isometric drawing as they do not appear as their true sizes. That this is so can be seen from Figure 1(b). All the angles on the cube are right angles but they are represented by angles of 60° and 120°, never by a right angle. Lines positioned by angles are drawn by fixing their ends by ordinates which are isometric lines. The part of the orthographic view which shows the line located by the angle is drawn, and the ordinates are transferred to the isometric view. The method is shown in Example 2.

ISOMETRIC PROJECTION

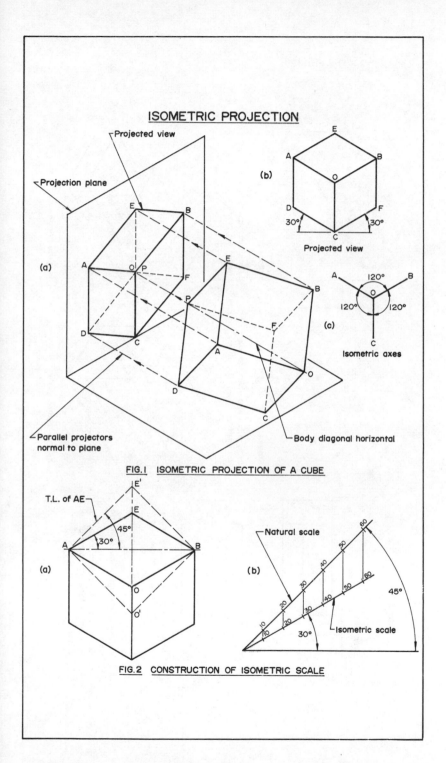

FIG.1 ISOMETRIC PROJECTION OF A CUBE

FIG.2 CONSTRUCTION OF ISOMETRIC SCALE

ISOMETRIC PROJECTION

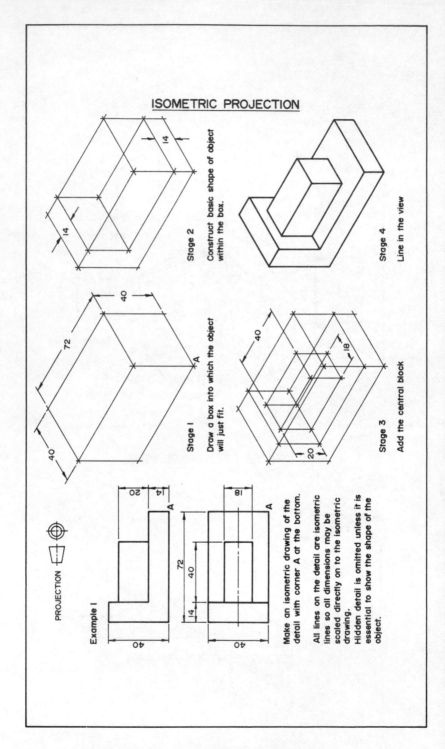

Stage 2

Construct basic shape of object within the box.

Stage 4

Line in the view

Stage 1

Draw a box into which the object will just fit.

Stage 3

Add the central block

PROJECTION

Example I

Make an isometric drawing of the detail with corner A at the bottom.

All lines on the detail are isometric lines so all dimensions may be scaled directly on to the isometric drawing.

Hidden detail is omitted unless it is essential to show the shape of the object.

152

ISOMETRIC PROJECTION

Stage 2

Construct 30° angle true size and find dimension X.

Stage 4

Line in the view

Stage 1

Draw box and construct basic shape as in Example I.

Stage 3

Transfer dimension X to isometric drawing and draw 30° angle.

PROJECTION

Example 2

Make an isometric drawing of the detail with corner A at the bottom.

The 30° angle is constructed using dimension X. The rest of the view is drawn as in Example I.

ISOMETRIC PROJECTION

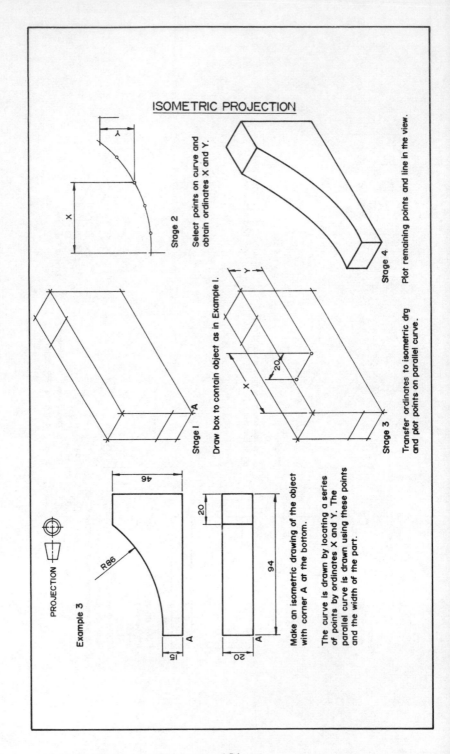

Stage 2

Select points on curve and obtain ordinates X and Y.

Stage 1

Draw box to contain object as in Example I.

Stage 3

Transfer ordinates to isometric drg and plot points on parallel curve.

Stage 4

Plot remaining points and line in the view.

Example 3

PROJECTION

R86

46

15

20

94

20

A

A

Make an isometric drawing of the object with corner A at the bottom.

The curve is drawn by locating a series of points by ordinates X and Y. The parallel curve is drawn using these points and the width of the part.

154

ISOMETRIC PROJECTION

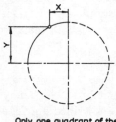 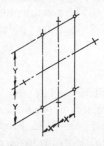 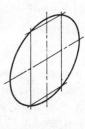

Only one quadrant of the
circle need be drawn

FIG.3   CIRCLE CONSTRUCTION BY ORDINATES

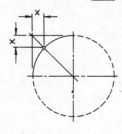 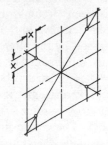 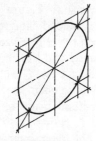

FIG.4   CIRCLE CONSTRUCTION BY CIRCUMSCRIBING SQUARE METHOD

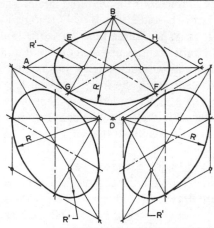 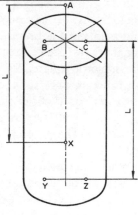

FIG.5   APPROXIMATE CIRCLE CONSTRUCTION

155

Objects with a curved profile

These are drawn by selecting a number of points on the curved profile in the orthographic view, and transferring ordinates of each point to the isometric view, as shown in Example 3. Usually a parallel curve is also required. This can be obtained by drawing parallel lines through the points on the first curve in the appropriate direction, and marking off on them the width of the detail. Too many points should not be taken or the work becomes tedious, but too few points will give an inaccurate profile.

Circles in isometric projection

Any circle on an isometric plane will appear as an ellipse which may be drawn in several ways. Ordinates may be used as described above, and this is illustrated in Figure 3 on page 155. From one quadrant of the full size circle, dimensions X and Y are transferred to the isometric view. The ellipse is completed with a french curve.

The circle in Figure 4 has been circumscribed by a square with the diagonals added. The square is redrawn in the isometric view and eight points on the ellipse are obtained by the intersection of the centre lines and square, and the diagonals and circle. The ordinates X fix the points on the diagonals. The ellipse is completed as before with a french curve.

The ellipse occurs so frequently on isometric drawings that an approximate method using circular arcs is often used to draw it. One such method is shown in Figure 5. The centre lines EF and GH are first drawn and their intersection made the centre of an isometric square ABCD with sides equal to the diameter of the required circle. The long diagonal AC of the square is drawn, and either B or D is joined to the mid-points of the opposite sides. Where these lines cross the long diagonal are two arc centres. The other centres are the corners B and D of the isometric square. The construction is facilitated if it is noted that all construction lines for the arc centres on the long diagonal are either horizontal or at 60° of the horizontal. This method is a special case of the general construction described in the following chapter on oblique projection. The general construction is illustrated in Figure 7 on page 180.

When drawing a cylinder as in Figure 5, the above construction need only be performed for one end. To obtain the centres for the other, lines parallel to the cylinder axis are drawn through centres, A, B and C, and on them the length L of the cylinder is marked off. This fixes the centres X, Y and Z. The same procedure may be used to draw a hole, if both ends are visible.

If a semicircle or quadrant is needed, only enough of the construction should be drawn to find the required centre or centres. For other fractions of a circle the ordinate method must be used, since the approximate method deviates considerably from a true ellipse at several points on the curve. This means that the approximate curve will not blend with other curves which are tangential to it at these points.

There is no doubt that the time taken to draw isometric circles is a serious disadvantage of the projection system. When a large number of circles are required on the drawing the work becomes tedious in the extreme, whichever method described above is used. In these circumstances ellipse templates are by far the simplest and quickest way of drawing the isometric circles, since the only construction needed is to lay out the centre lines of the circles. These are aligned with the corresponding centre lines on the template and the ellipse traced round. It is obviously impossible to provide templates for every possible ellipse, but by choosing a large template and leaning the pencil outwards from the hole, a satisfactory curve can be drawn. Alternatively the template can be shifted slightly as each quarter of the ellipse is drawn. All the ellipses in the solution to Example 8 on page 163 were drawn using templates.

Alternative positions of the isometric axes

Some details are shown to their best advantage by varying the position of the isometric axes. In Example 4 the object when in use is viewed from below, and the position of the axes is chosen to produce a view as seen from below. Occasionally long objects may be best drawn with their axes horizontal, as in Example 5.

Isometric sections

The internal features of an object may be shown by using sections, as illustrated in Examples 6 and 7. Half or full sections may be used depending on the object, and occasionally local sections are useful. When drawing a half section it is best to block in lightly the complete isometric view and then to remove the front quarter to expose the interior. With full sections the plane of the section should be constructed first, and then the remainder of the object drawn behind it. Section lines are best drawn at 60° to the horizontal, and on half sections slope in opposite directions on the cut surfaces.

Exploded assemblies

These are used extensively in instruction books, maintenance and repair manuals and similar technical publications. They show the parts which comprise the assembly in their correct relative positions, but spread out along the centre lines so that they can be easily identified. Where possible the parts should be positioned so that they do not overlap, but with assemblies having many parts this cannot always be done. For large assemblies the parts list appears on a separate sheet. An assembly in exploded form is shown in Example 8.

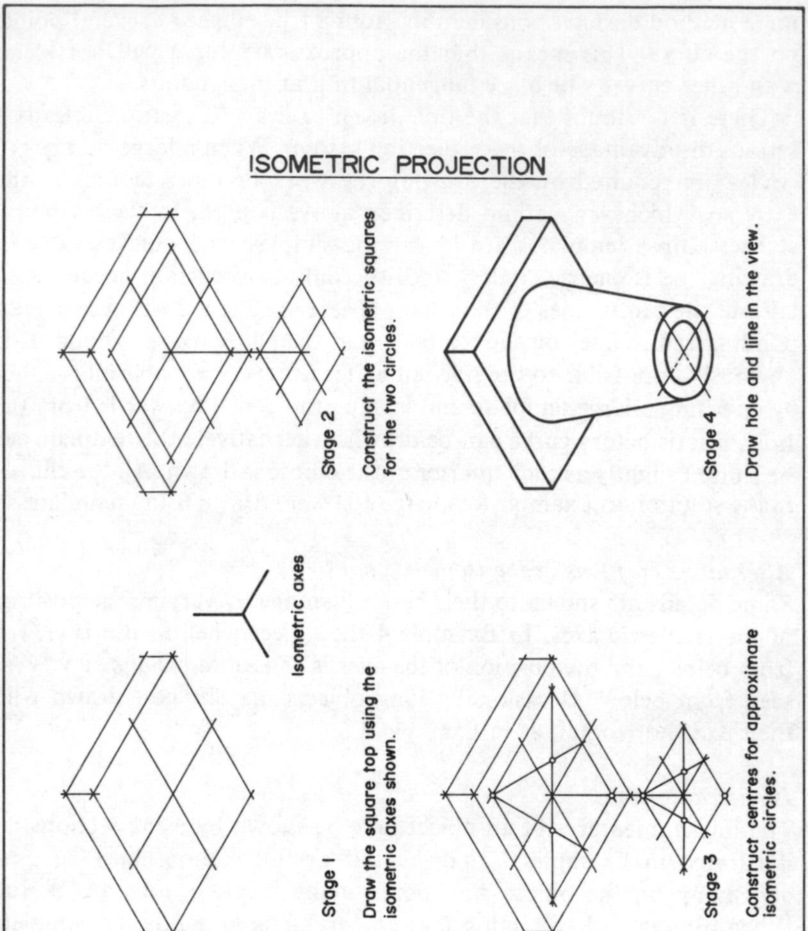

ISOMETRIC PROJECTION

Stage 2

Construct the isometric squares for the two circles.

Stage 4

Draw hole and line in the view.

Isometric axes

Stage I

Draw the square top using the isometric axes shown.

Stage 3

Construct centres for approximate isometric circles.

PROJECTION

Example 4

Ø15

Ø54

Ø40

Ø28

10

44

Make an isometric drawing of the component viewing it from below.

158

ISOMETRIC PROJECTION

Isometric axes

Stage 1

Draw box using the isometric axes shown.

Stage 2

Draw square end and construction for isometric circles.

Stage 3

Draw isometric circles and line in the view.

PROJECTION

Example 5

Ø24

□30

Ø52

32

120

160

Make an isometric drawing of the component with the axis horizontal.

ISOMETRIC PROJECTION

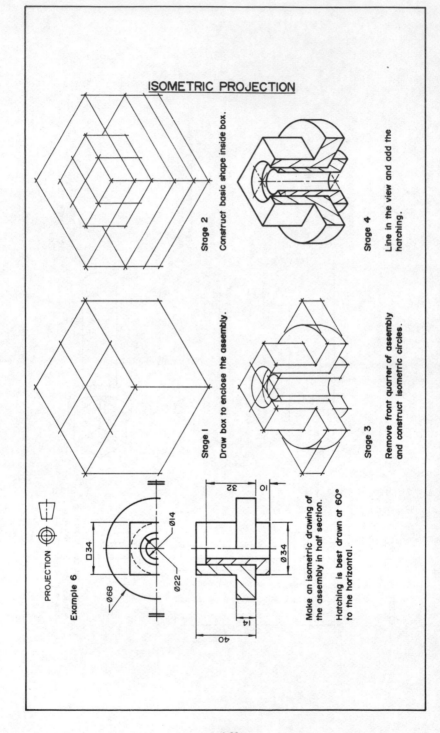

PROJECTION

Example 6

∅68

□34

∅14

∅22

Make an isometric drawing of
the assembly in half section.

Hatching is best drawn at 60°
to the horizontal.

32

10

∅34

14

40

Stage I

Draw box to enclose the assembly.

Stage 2

Construct basic shape inside box.

Stage 3

Remove front quarter of assembly
and construct isometric circles.

Stage 4

Line in the view and add the
hatching.

160

ISOMETRIC PROJECTION

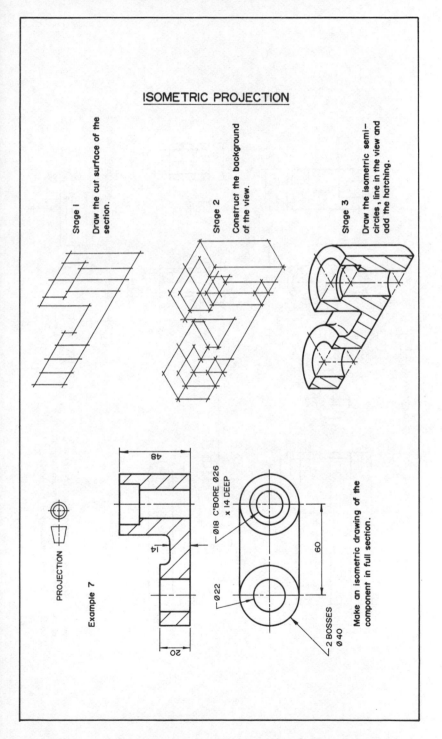

Stage 1

Draw the cut surface of the section.

Stage 2

Construct the background of the view.

Stage 3

Draw the isometric semi-circles, line in the view and add the hatching.

PROJECTION

Example 7

48

Ø18 C'BORE Ø26 x 14 DEEP

14

Ø22

60

2 BOSSES Ø40

20

Make an isometric drawing of the component in full section.

161

ISOMETRIC PROJECTION

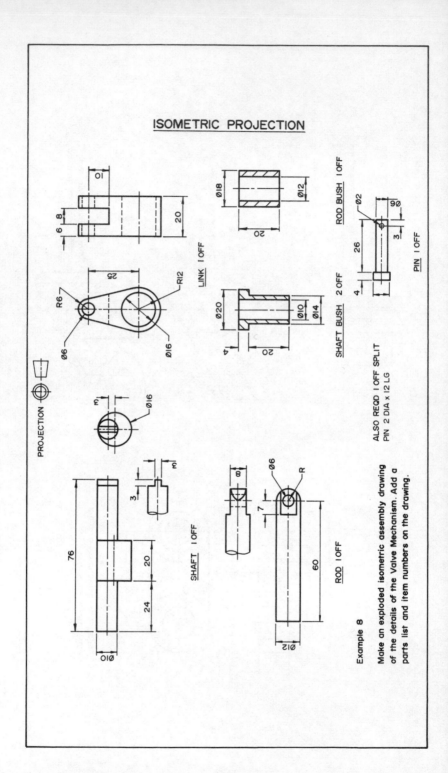

LINK 1 OFF

ROD BUSH 1 OFF

PIN 1 OFF

SHAFT BUSH 2 OFF

ALSO REQD 1 OFF SPLIT
PIN 2 DIA x 12 LG

PROJECTION

SHAFT 1 OFF

ROD 1 OFF

Example 8

Make an exploded isometric assembly drawing of the details of the Valve Mechanism. Add a parts list and item numbers on the drawing.

162

ISOMETRIC PROJECTION

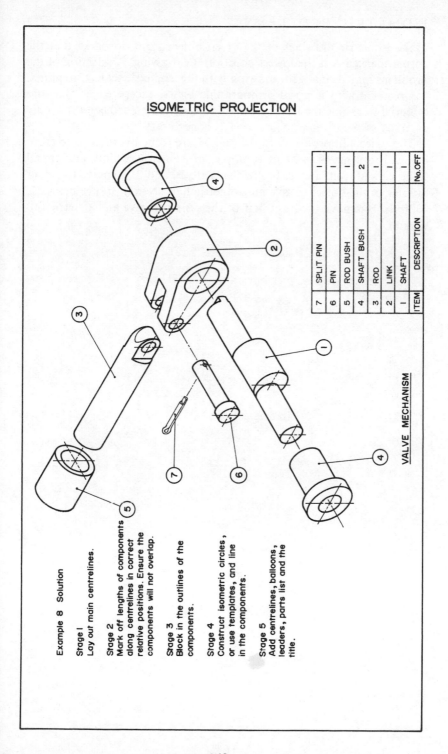

VALVE MECHANISM

ITEM	DESCRIPTION	No.OFF
7	SPLIT PIN	—
6	PIN	—
5	ROD BUSH	—
4	SHAFT BUSH	2
3	ROD	—
2	LINK	—
1	SHAFT	—

Example 8 Solution

Stage 1
Lay out main centrelines.

Stage 2
Mark off lengths of components along centrelines in correct relative positions. Ensure the components will not overlap.

Stage 3
Block in the outlines of the components.

Stage 4
Construct isometric circles, or use templates, and line in the components.

Stage 5
Add centrelines, balloons, leaders, parts list and the title.

163

ISOMETRIC PROBLEMS

Make isometric drawings of the given objects, positioned so that the corner marked A is the lowest point on the drawing. A selection of the problems may be made for drawing with the isometric scale, if required.

Do not draw the given orthographic views, except where they are essential to enable the isometric view to be completed. Then draw only as much of the orthographic views as is necessary.

The objects in problems 15, 16 and 18 are to be drawn as seen from below. Problems 19 to 24 are examples of isometric sections and are all to be viewed from above. In problems 19, 20 and 21, remove the front quarter of the object to give an isometric half section. In problems 22, 23 and 24, remove the front half of the object to give an isometric full section.

ISOMETRIC PROBLEMS

PROJECTION

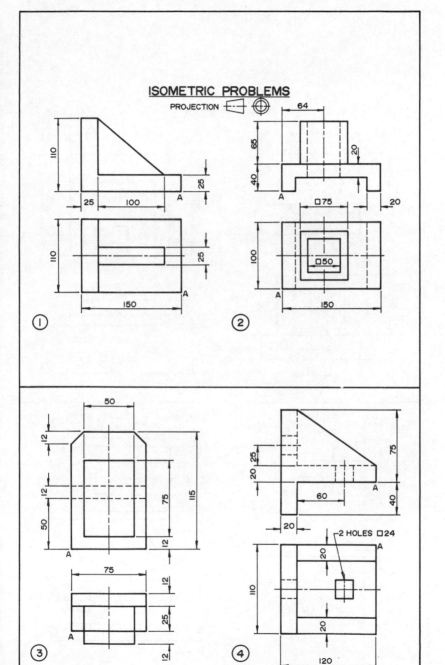

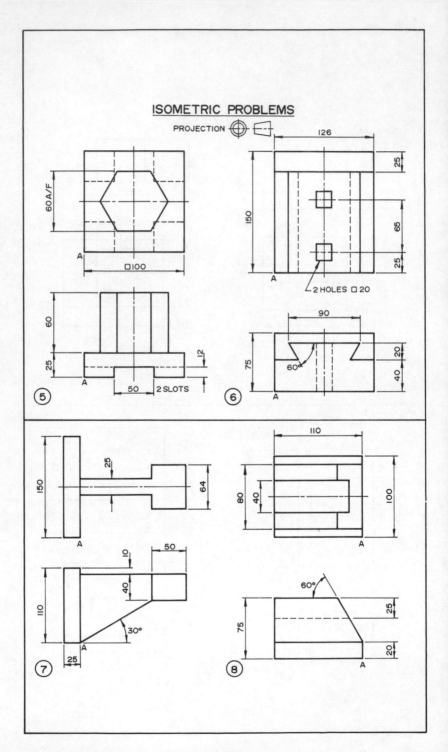

ISOMETRIC PROBLEMS

PROJECTION ⊕ ◁

5

60 A/F

□100

60

25

A

50   2 SLOTS

12

6

126

25

150

65

25

A

2 HOLES □ 20

90

75

60°

20

40

A

7

150

25

64

A

110

10   50

40

30°

25

A

8

110

80

40

100

A

60°

75

25

20

A

166

ISOMETRIC PROBLEMS

PROJECTION ⊏⊐ ⊕

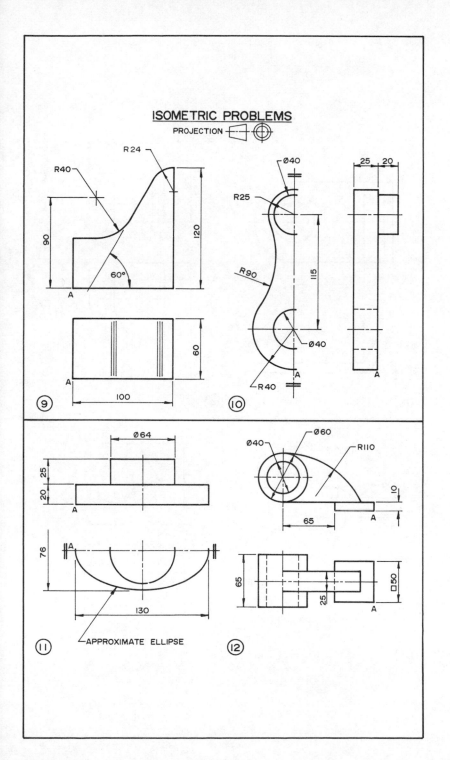

R24

R40

90

120

60°

A

100

60

⑨

Ø40

R25

R90

115

Ø40

A

R40

25 20

A

⑩

Ø64

25

20

A

76

130

⑪

APPROXIMATE ELLIPSE

Ø40

Ø60

R110

10

65

A

65

25

□50

A

⑫

167

ISOMETRIC PROBLEMS

PROJECTION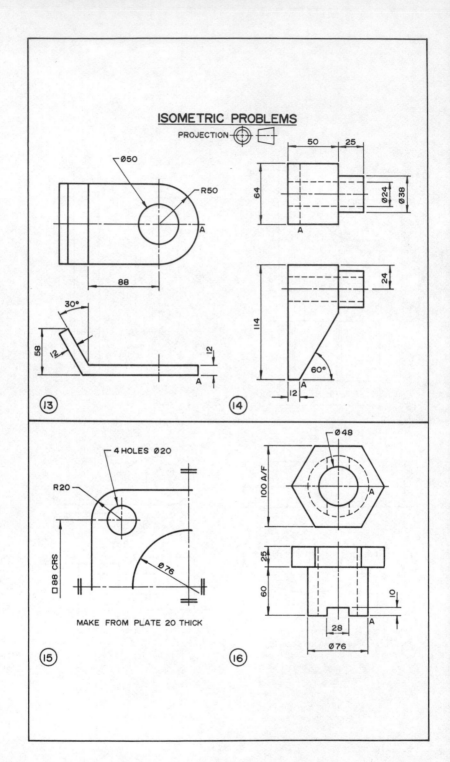

Ø50
R50
A
88
30°
58
12
12
A
⑬

50 25
64
Ø24 Ø38
A
114
24
60°
A
12
⑭

4 HOLES Ø20
R20
□ 88 CRS
Ø76
MAKE FROM PLATE 20 THICK
⑮

Ø48
100 A/F
A
25
60
10
28
Ø76
A
⑯

168

ISOMETRIC PROBLEMS

PROJECTION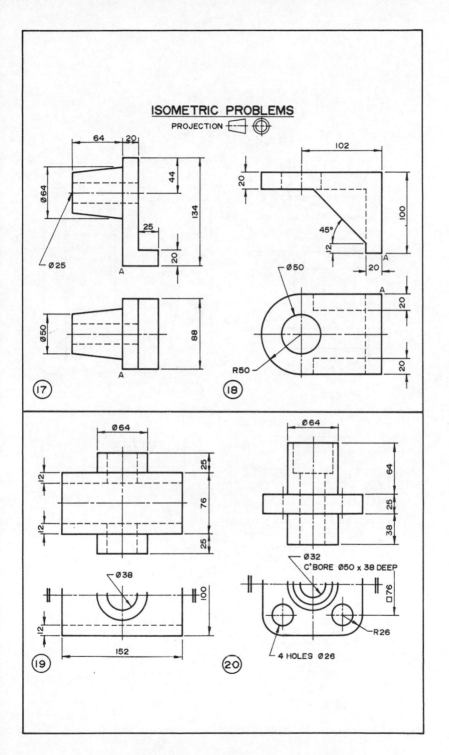

17

⌀64
64 20
44
134
25
20
⌀25
A

⌀50
88
A

18

102
20
100
45°
12
20
A

⌀50
20
20
R50
A

19

⌀64
25
76
25
12
12

⌀38
100
12
152

20

⌀64
64
25
38

⌀32
C'BORE ⌀50 x 38 DEEP
□76
R26
4 HOLES ⌀26

169

ISOMETRIC PROBLEMS

PROJECTION

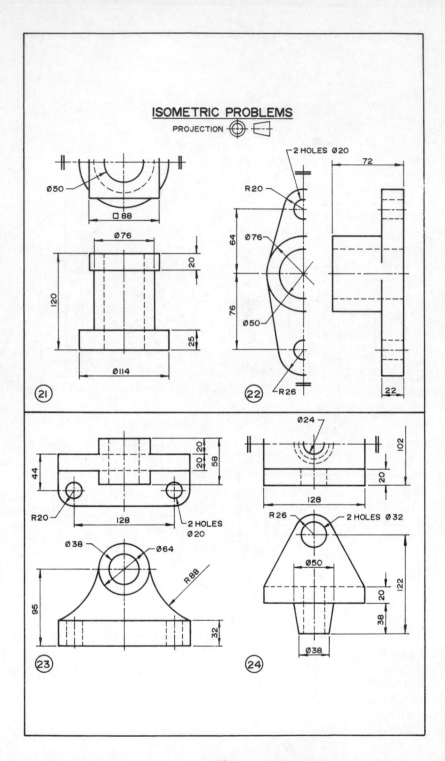

21

22

23

24

15

OBLIQUE PROJECTION

OBLIQUE projection, or more correctly, oblique parallel projection, is, like isometric projection, a system for producing a pictorial view of an object. In isometric projection the projectors from the object to the projection plane are parallel and normal to the plane. In oblique parallel projection they are parallel but oblique to the plane, and the object is positioned with a face parallel to the plane. This is in contrast to isometric projection where no face of the object is parallel to the projection plane. Figures 1 and 2 illustrate the principles of the system.

Figure 1(a) shows a line AB perpendicular to a plane. Using parallel projectors the projection of AB on the plane may vary from a point to a line of infinite length. Thus, if the projectors are perpendicular to the plane, they will coincide with AB producing a point view. If the projectors are parallel to the plane, a projection of infinite length will result at an infinite distance along the plane. Between these limits any length of projection may be obtained, but if the angle made by the projectors with the plane is 45° the projection will be true length, since triangles OAa^1 and OBb^1 are similar. It should be noted that the projections of lines which are parallel to the plane will always be true lengths, regardless of the angle which the projectors make with the plane, provided that the projectors are parallel.

In Figure 1(b) the right angled triangle OBb^1, which contains the line AB and its true length projection a^1b^1, has been revolved round OB. It is immediately apparent that the projections a^1b^1, a^2b^2 and a^3b^3 are all the same length, that is, the true length of AB. Therefore, provided the true angle between the projectors and the projection plane is 45°, the angle between the plane containing the projectors and the horizontal plane may have any value, for a true length projection of the line to be obtained.

A square ABCD, perpendicular to the projection plane, is shown in Figure 2(a). The square is projected on to the plane by parallel projectors making an angle of 45° with the plane, the projectors lying in a plane at 30° to the horizontal. From the foregoing discussion it can be seen that all sides of the projected view are true lengths. The shape, however, is not true, being a rhombus instead of a square.

In Figure 2(b) a cube is shown arranged with a pair of faces parallel to the projection plane. It is projected on to the plane using the same system of parallel projectors as in Figure 2(a). Thus, all edges of the view are true lengths, and the faces BDEF and ACGH are true shape, since they are parallel to the projection plane. It follows also that the circle on plane BDEF will project as a true circle, although an observer viewing it along the projectors will see an ellipse. This is a curious state of affairs but it illustrates the principal advantage of oblique parallel projection over isometric projection. This is, that it is often possible to put the object to be drawn in a position where circles and circular arcs may be drawn with compasses. It can also be seen from Figure 2(b) that the oblique projection of an object may, in general, be drawn directly, without first drawing the orthographic views. The three axes, ab, db and eb, are drawn and on them the view is built up. The axis ab is called the cross or receding axis.

Cavalier and cabinet projection

The oblique projection of the cube in Figure 2(b) has been redrawn in the left-hand view of Figure 3. Although this is a true projection the cube appears too long. This is caused by no allowance being made for perspective or the apparent convergence of receding parallel lines. If the scale on the receding axis is reduced the distortion becomes less apparent, and scales of $\frac{3}{4}$ full size and $\frac{1}{2}$ full size are commonly used. The view will still be a true projection whatever the scale used on the receding axis, since the scale may be changed merely by altering the angle which the projectors make with the projection plane, as shown in Figure 1(a). If full size is used on the receding axis the view is said to be in cavalier projection. The name refers to its use in the past to draw fortifications, all dimensions of which needed to be scaled directly from the drawing. A scale of $\frac{1}{2}$ full size on the receding axis results in cabinet projection. This was used to draw furniture (cf. 'cabinet maker'). Here the front face of the piece is generally more important and complicated than its sides, and the depth is usually much less than the length or height. If cavalier projection is used, the depth will appear to be much bigger than it actually is.

Variation of the angle of the receding axis

Different faces of the object may be shown in oblique projection by altering the angle which the receding axis makes with the horizontal. True projections will again result, since the angle of the receding axis depends on the angle which the parallel projectors make with the horizontal, and, as shown above, this angle may have any value.

OBLIQUE PROJECTION

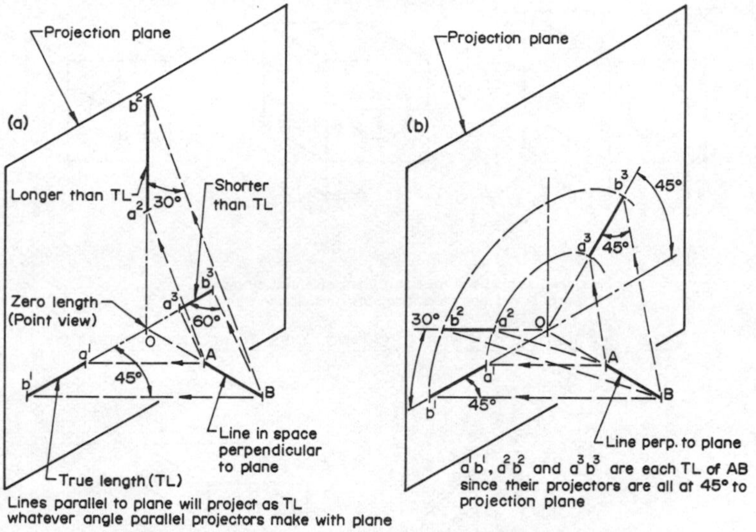

(a)
Projection plane
Longer than TL
Shorter than TL
30°
Zero length (Point view)
60°
45°
Line in space perpendicular to plane
True length (TL)
Lines parallel to plane will project as TL whatever angle parallel projectors make with plane

(b)
Projection plane
45°
45°
30°
45°
Line perp. to plane
a^1b^1, a^2b^2 and a^3b^3 are each TL of AB since their projectors are all at 45° to projection plane

FIG.1 TRUE LENGTHS OF LINES IN OBLIQUE PARALLEL PROJECTION

(a)
Projection plane
TL of AB
30°
45°
All sides of projection are TL. Shape is not true
AC and BD are parallel to projection plane so their projections are TL

(b)
Projection plane
All sides of projection are TL
30°
True shape circle
Parallel projectors at 45° to projection plane
Faces ACGH and BDEF are parallel to projection plane and so project as true size and shape

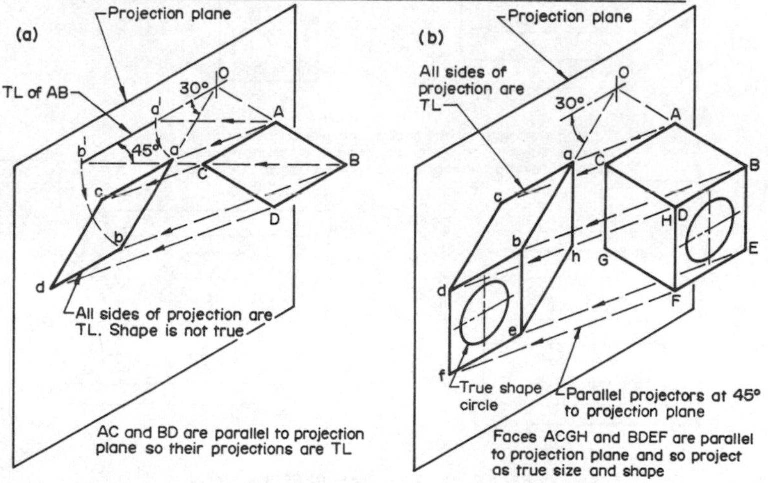

FIG.2 OBLIQUE PARALLEL PROJECTION OF SQUARE AND CUBE

OBLIQUE PROJECTION

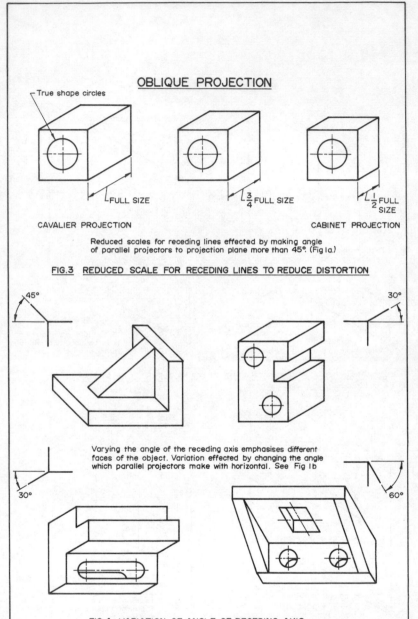

True shape circles

FULL SIZE

$\frac{3}{4}$ FULL SIZE

$\frac{1}{2}$ FULL SIZE

CAVALIER PROJECTION

CABINET PROJECTION

Reduced scales for receding lines effected by making angle
of parallel projectors to projection plane more than 45°. (Fig 1a.)

FIG.3 REDUCED SCALE FOR RECEDING LINES TO REDUCE DISTORTION

45°

30°

Varying the angle of the receding axis emphasises different
faces of the object. Variation effected by changing the angle
which parallel projectors make with horizontal. See Fig 1b

30°

60°

FIG.4 VARIATION OF ANGLE OF RECEDING AXIS

OBLIQUE PROJECTION

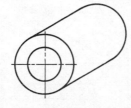

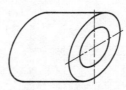

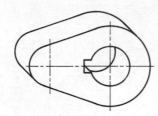

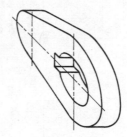

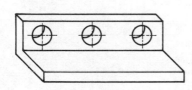

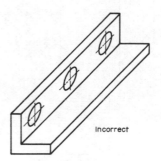

Correct. Minimum distortion,
circles and arcs drawn with
compasses

Incorrect

Place the face of the object containing circles or arcs
parallel to the projection plane.

Place the longest dimension of the object parallel to
the projection plane.

FIG.5   POSITIONING OBJECT RELATIVE TO PROJECTION PLANE

OBLIQUE PROJECTION

PROJECTION

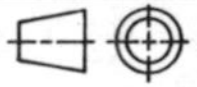

Example I

2 HOLES Ø10

54
40
60°
44
22
12
12
12

72
12
32

FULL SCALE

Stage I

Draw box to enclose object. Use receding axis angle to show angled face

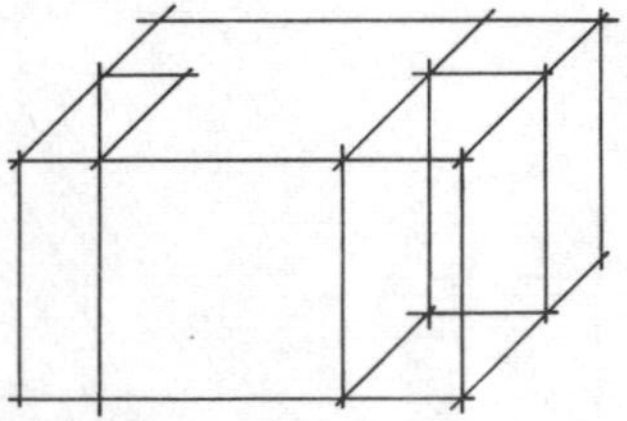

Stage 2

Construct basic shape of object within the box.

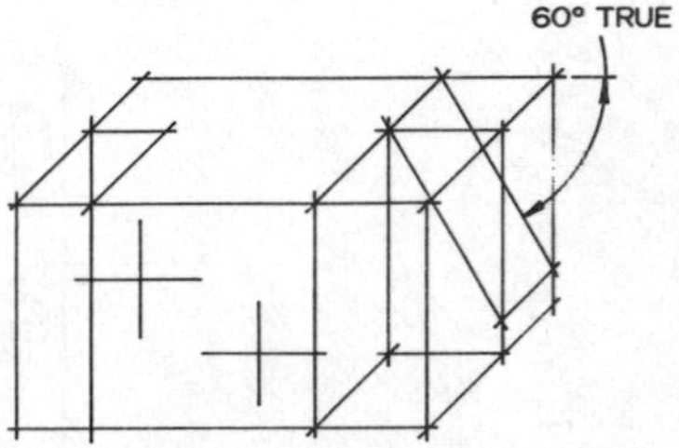

Stage 3

Construct angled face and lay out hole centres.

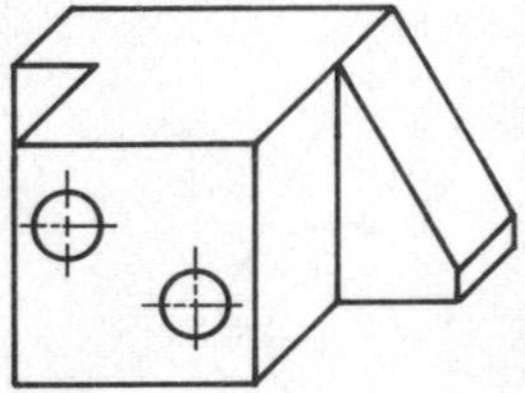

60° TRUE

Stage 4

Draw holes and line in the view.

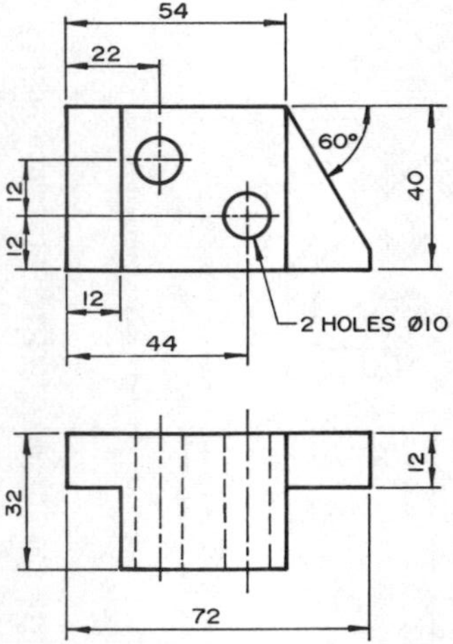

Make an oblique drawing of the component in cavalier projection. Select a suitable angle for the receding axis.

Do not show hidden detail unless it is essential to show the shape of the object.

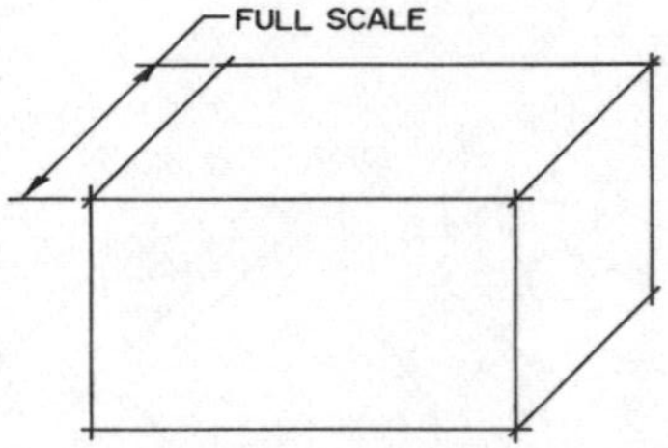

176

OBLIQUE PROJECTION

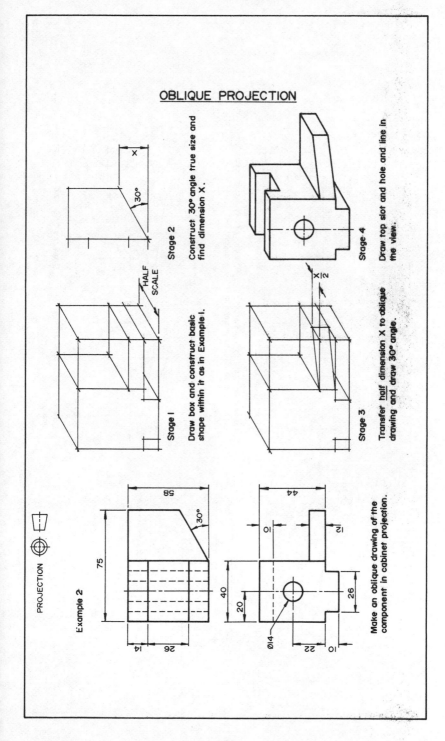

Stage 2

Construct 30° angle true size and find dimension X.

Stage 1

Draw box and construct basic shape within it as in Example 1.

Stage 3

Transfer half dimension X to oblique drawing and draw 30° angle.

Stage 4

Draw top slot and hole and line in the view.

PROJECTION

Example 2

Make an oblique drawing of the component in cabinet projection.

177

OBLIQUE PROJECTION

PROJECTION

Example 3

Ø22

R15

2 HOLES Ø10

R10

46

45°

52

Ø36

12

26

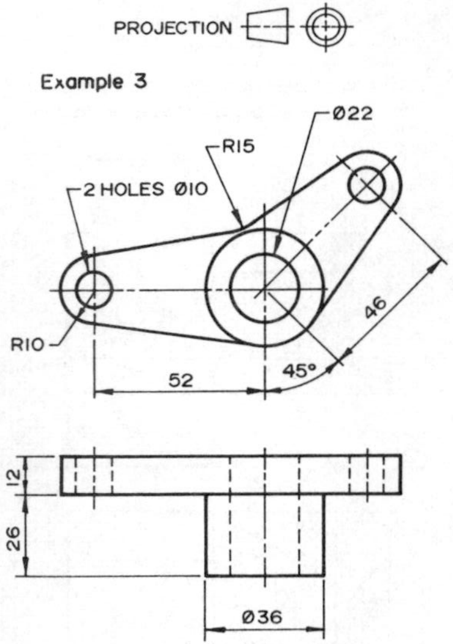

Make an oblique drawing of the lever using cavalier projection.

For accuracy it is best to locate all points of tangency.

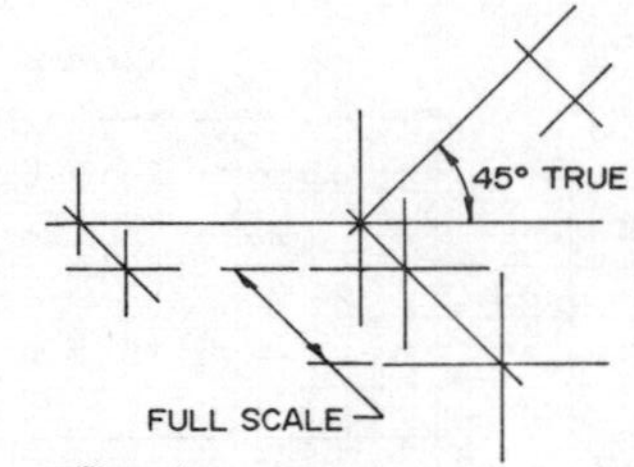

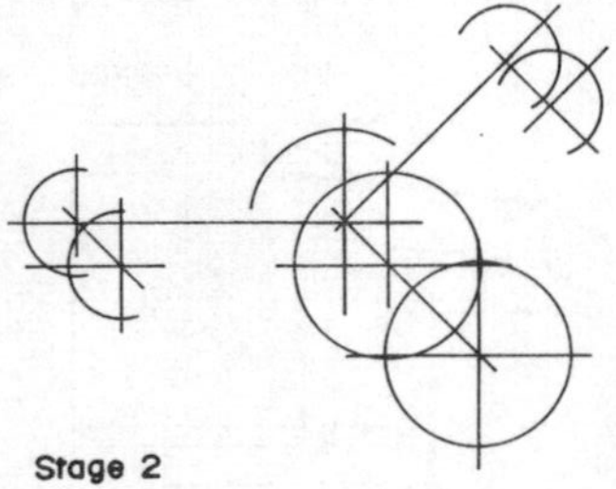

Stage 2

Draw circles and circle arcs for lever profile.

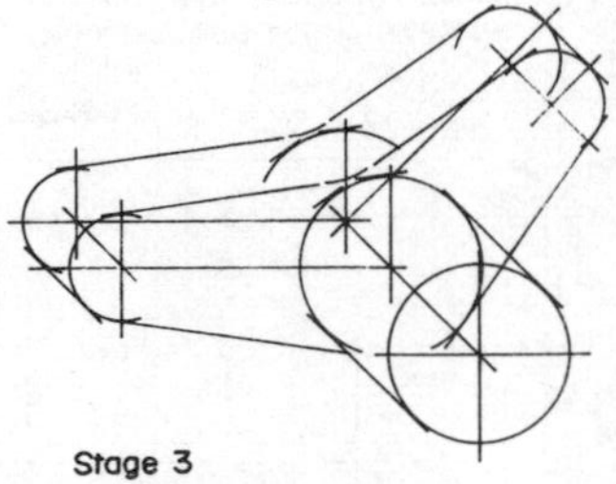

FULL SCALE

45° TRUE

Stage 1

Lay out the centre lines so that circles are parallel to projection plane.

Stage 3

Complete the profile with the straight lines.

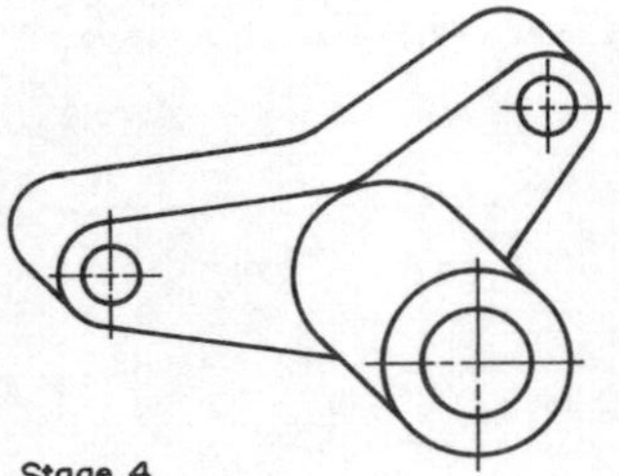

Stage 4

Draw the holes and line in the view.

178

Positioning the object relative to the projection plane

It can be seen, therefore, that whatever scale is used on the receding axis, and whatever angle it makes with the horizontal, the view will be a true projection. However, the distortion caused by drawing the receding lines parallel may be objectionable when the object is in some positions relative to the projection plane, as shown in Figure 5. To minimize the distortion, two rules should be observed. First, place the face of the object which contains circles or circular arcs parallel to the projection plane. Second, place the longest dimension of the object parallel to the projection plane. When these two rules conflict, as with long objects of circular cross-section, the first takes precedence.

The first rule has been applied in the solution to Example 1 on page 176. The faces containing the circular holes and the 60° angle have been positioned parallel to the projection plane so that they may be drawn true size and shape. The receding axis is drawn upwards to the right to show the 60° angle. This example also illustrates the use of a box which just encloses the object, as the first step in making the drawing. This technique is the same as that used in isometric projection.

Angles on a receding plane

In Example 2, the 30° angle does not lie in a parallel plane to the 14 mm diameter hole, so it will not appear as its true size in the oblique drawing. However, from a true size view, the dimension X can be transferred to the oblique drawing, but since cabinet projection is being used X must be halved during transfer.

Circles in oblique projection

Example 3 is of the type which is ideally suited for drawing in oblique projection. In isometric projection the circles and circular arcs would take much time to construct; in oblique projection they are drawn directly with compasses. Instead of drawing a box, it is more convenient with objects like this to draw first the centre lines and use them as a skeleton on which to complete the view.

When an object has circles or circular arcs on two planes at right angles, those on one plane will project as ellipses. Figure 6 shows the use of ordinates to draw such ellipses, and it will be seen that the method is the same as that used in isometric projection. When cabinet projection is used, it must be remembered that the ordinate laid off parallel to the receding axis is halved, as shown in the solution to Example. 4.

The approximate circle construction used in isometric projection can also be used to draw circles on a receding plane, and is illustrated in Figure 7. The four centres for the arcs lie at the intersections of the perpendicular bisectors of the sides of the circumscribing square. In

OBLIQUE PROJECTION

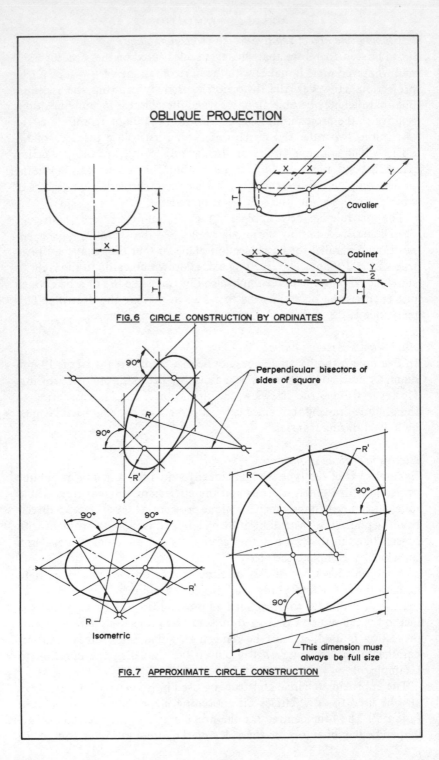

FIG.6   CIRCLE CONSTRUCTION BY ORDINATES

Cavalier

Cabinet

Perpendicular bisectors of
sides of square

Isometric

This dimension must
always be full size

FIG.7   APPROXIMATE CIRCLE CONSTRUCTION

180

OBLIQUE PROJECTION

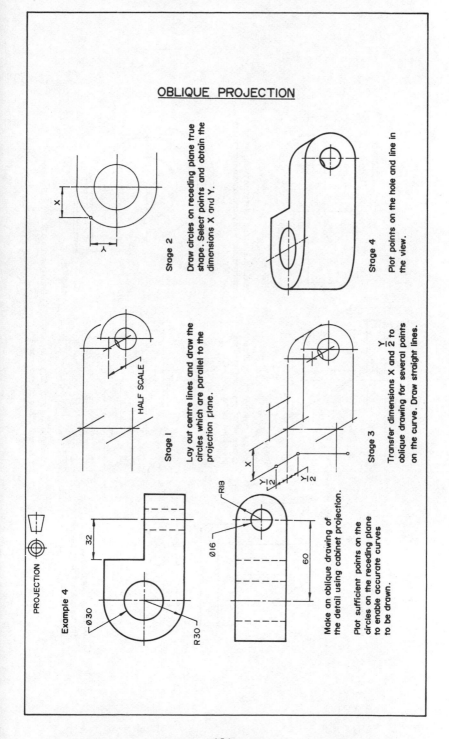

PROJECTION

Example 4

Ø30

32

R30

RI8

Ø16

60

Make an oblique drawing of the detail using cabinet projection.

Plot sufficient points on the circles on the receding plane to enable accurate curves to be drawn.

Stage 1

Lay out centre lines and draw the circles which are parallel to the projection plane.

HALF SCALE

Stage 2

Draw circles on receding plane true shape. Select points and obtain the dimensions X and Y.

X

Y

Stage 3

Transfer dimensions X and $\frac{Y}{2}$ to oblique drawing for several points on the curve. Draw straight lines.

X

$\frac{Y}{2}$

$\frac{Y}{2}$

Stage 4

Plot points on the hole and line in the view.

181

OBLIQUE PROJECTION

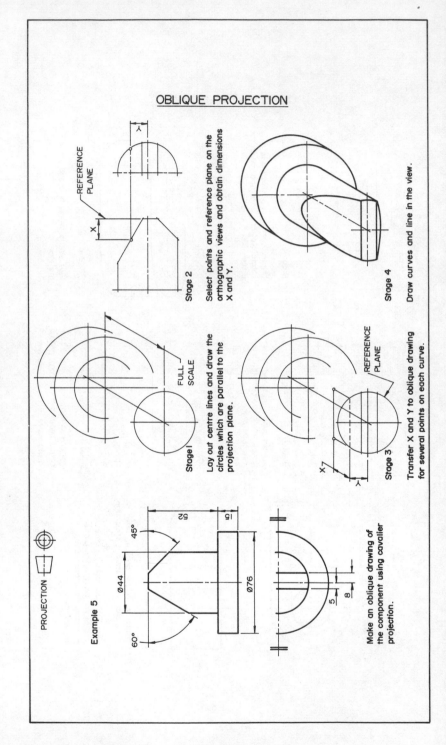

Stage 2

Select points and reference plane on the orthographic views and obtain dimensions X and Y.

Stage 4

Draw curves and line in the view.

Stage 1

Lay out centre lines and draw the circles which are parallel to the projection plane.

FULL SCALE

Stage 3

Transfer X and Y to oblique drawing for several points on each curve.

REFERENCE PLANE

Example 5

Make an oblique drawing of the component using cavalier projection.

45°
60°
52
15
Ø44
Ø76
5
8

PROJECTION

182

OBLIQUE PROJECTION

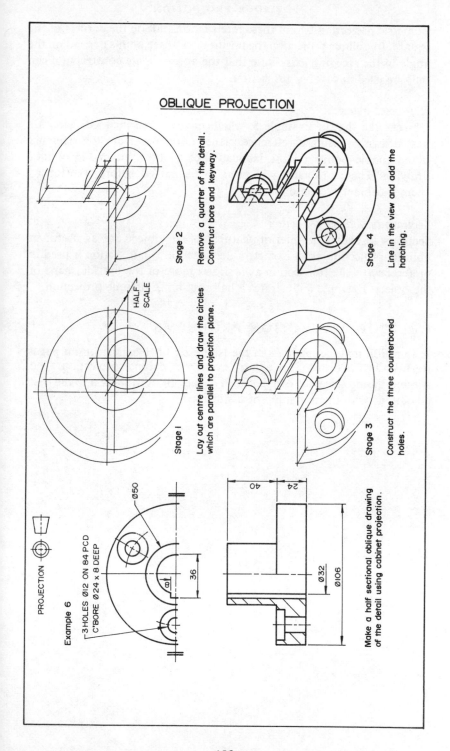

Stage 2

Remove a quarter of the detail. Construct bore and keyway.

Stage 4

Line in the view and add the hatching.

HALF SCALE

Stage 1

Lay out centre lines and draw the circles which are parallel to projection plane.

Stage 3

Construct the three counterbored holes.

PROJECTION

Example 6

3HOLES Ø12 ON 84 PCD
C'BORE Ø24 x 8 DEEP

Ø50

36

18

40

24

Ø32

Ø106

Make a half sectional oblique drawing of the detail using cabinet projection.

183

isometric projection two of these intersections fall on the corners of the square. In oblique projection the positions of these points depend on the angle of the receding axis. Note that the approximate construction can only be used in cavalier projection.

Reference planes

Objects like that in Example 5, which have curved inclined faces, are best drawn by using a reference plane. Ordinates from the reference plane enable points on the inclined face to be fixed in the oblique drawing. These ordinates will need to be halved before transferring them if cabinet projection is used.

Sections in oblique projection

Sections to show the interior features of an object are as useful in oblique projection as in isometric. In general, a half section is used in preference to the full section as it shows more of the outside shape of the object. Example 6 illustrates a half section in oblique projection.

OBLIQUE PROJECTION PROBLEMS

A selection may be made from the isometric problems given on pages 165 to 170 for drawing in either cavalier or cabinet projection. The student should select his own position of the object to show it to the best advantage with the minimum of distortion.

16

TECHNICAL SKETCHING

EVERY engineer should acquire the ability to produce clear, neat and accurate sketches, since they have so many applications. Most new ideas first take shape as freehand sketches, which are refined and modified until they are in a suitable form from which a drawing made with instruments can be produced. In a technical discussion a rapid sketch will often supplement the spoken word, resulting in clearer and quicker communication. Instructions on the shop floor or in the field can be conveyed unambiguously and rapidly with sketches. Alternative solutions to a problem can be worked out in sketch form before the best solution is chosen, the sketches providing a valuable permanent record of the reasons for the choice of the final solution. Information for the replacement of broken parts and similar one-off jobs is provided more economically by a sketch than by an instrument drawing. Finally, the detail draughtsman may use sketches before starting the instrument drawing, to decide such things as the number of views required and their layout, how much space to allow for dimensions, and so on.

It is unfortunate that in everyday speech, the term sketch implies something which is rough and incomplete, needing artistic talent for its production. Technical sketches made to clarify spoken descriptions in a discussion may be unfinished, but they should not be made carelessly or they will not achieve their object. Sketches from which a component is to be made should be as complete as a drawing made with instruments. To make a good technical sketch does not require any artistic talent. All that is needed is the mastery and application of a few simple techniques.

Equipment

The only equipment required for sketching is a pencil, an eraser and some paper. The pencil should be fairly soft, such as an HB or F grade, and sharpened to a conical, rounded point. The eraser should be an ordinary soft pencil eraser. A4 size paper, printed in 5 mm or 10 mm squares, is the best for beginner and expert alike, since the printed lines take the place of the T square, set squares and scale. Some people find it difficult to sketch on a pad because of its thickness. For them a clip board, to take one sheet only, is recommended. If the printed lines on the squared paper are thought objectionable in the finished sketch, plain typing paper over the squared paper will enable the lines to be used without them appearing on the finished work.

Technique

Do not use paper larger than A4, or exceptionally A3, and do not fix it to a drawing board. So that lines may be drawn easily and naturally, it is essential to be able to turn the paper round as the work proceeds, and large sheets prevent this being done conveniently. For sketching, a drawing board is an encumbrance, since it cannot be carried easily and there will often be nowhere to rest it in the workshop or in the field. All lines must be drawn freehand and the temptation, met by most beginners, to use a straight edge and compasses must be firmly resisted. Such 'directed sketches' as they are sometimes called, lose the advantage of rapid execution inherent in freehand work. Also, many sketches in industry must be made with the pad or clip board on one's knees, or while standing, and in these conditions no drawing instruments can be used.

For most people the easiest and most natural way to sketch a straight line is to work horizontally from left to right. Since the paper is not fixed to a drawing board when sketching, all straight lines may be drawn in this way, the paper being turned to bring the line horizontal. Put the pencil point on one end of the line, look at the other end, not the pencil point, and draw the line. Do not retrace the line or use short overlapping strokes as this will make it woolly and indistinct. Finished lines must be hard and clean. Short lines may be drawn in one movement of the pencil, using a wrist or arm motion rather than finger movement. For longer lines use several such strokes, leaving small gaps rather than trying to overlap them. Do not strive to imitate the straightness of mechancially drawn lines. Slight ripples and deviations from straightness are an essential part of the style of freehand sketching. However, lines must start and stop at exactly the right places and the eraser should be used to correct lines drawn too long. In instrument drawing the eraser should be used as little as possbile. In freehand sketching it is an essential tool.

The distinction between thick and thin lines must be preserved in a sketch, and the easiest way to achieve this is to re-point the pencil lead before drawing centre lines, dimension lines and hidden detail lines. Construction lines should be very faint and dimmed out with the eraser when the construction is complete.

Angles in a sketch may be approximated quite closely by using the vertical and horizontal lines on the squared paper. 45° angles are easily obtained by drawing diagonals through the squares in the required direction. 30° and 60° angles can be estimated by dividing a right angle into three parts by eye, and the same technique may be used for other angles as they occur.

Beginners approach the sketching of circles and circular arcs with considerable apprehension, thinking that it is in this area, more than any other, that artistic ability is required. This is quite wrong. Acceptable circles and arcs of any size may be sketched by applying simple constructions to the geometry of the circle. Small circles and arcs are easily sketched without any preliminary construction. For larger circles first sketch the centre lines and two 45° radial lines. Using the squares as a guide, mark off the estimated radius on the four lines, thus obtaining eight points on the circle. Through those points sketch light arcs, and extend them to form the complete circle. Remember that the arcs must cross the lines at right angles. Dim out the construction and darken the circle. In all circle sketching it is easier to work with the pencil on the inside of the curve, rather than on the outside.

Another method is to mark the radius of the circle on the edge of a piece of scrap paper and use this to position as many points as desired from the centre. Complete the circle with a dark line through the points. A third technique is to use the hand as a compass. Fix the position of the centre and place the tip of the little finger on it. Adjust the pencil in the hand to the required radius, and keeping the hand rigid, rotate the paper carefully with the other hand. A modification of this method is to use two pencils as a compass. Both pencils are held crossed in one hand, their points being the required radius apart. Place one point on the centre and revolve the paper under the other, which will sketch the circle. It is best with these methods to draw a faint line which can be darkened with the pencil held normally.

When sketching arcs tangential to straight lines or other arcs, construct the points of tangency approximately by using the same geometrical constructions in the sketch as would be employed if the drawing was made with instruments.

Scale

The scale to which a sketch is made is usually of no importance, but the features of the object must be shown in their correct proportions. The printed squares on the paper provide a simple means of keeping the proportions accurate, but distances marked on the edge of a piece of scrap paper may also be used. Always begin a sketch by fixing the overall sizes, working from them to the sizes of the smaller features. Choose as large a scale as possible, but if dimensions are required allow enough space for them to be inserted without crowding. If several orthographic views are to be sketched, corresponding dimensions on different views must agree or the readability of the sketch will be impaired.

Isometric and oblique sketches

Sketches in oblique projection with the receding axis at 45° may be rapidly drawn on squared paper, the receding axis following the diagonals of the squares. Special isometric paper which has printed lines in the directions of the axis may be obtained, and this facilitates the production of isometric sketches. For sketches in either pictorial projection establish the enclosing box first, exactly as if instruments are being used. To sketch isometric circles start with the centre lines and circumscribing isometric square. Then draw the two large arcs and complete with the small arcs. Bear in mind that the arcs must cross the centre lines at right angles or the circle will be distorted. The sketching of ellipses in isometric projection is made much easier if it is remembered that the major axis is always at right angles to the axis of the cylinder or hole, the major and minor axes are at right angles, and the minor axis coincides with cylinder or hole axis.

SKETCHING PROBLEMS

Use A4 squared paper for these problems, with squares of 5 mm or 10 mm side. Dimensions are approximate only but proportions should be maintained as accurately as possible.

1 Sketch sets of six parallel lines (a) 40 mm long (b) 80 mm long (c) 120 mm long (d) 200 mm long, with one square between pairs of lines.

2 Sketch two sets of six parallel lines 60 mm long and 10 mm apart, each set sloping at 45° in opposite directions.

3 Make sketches of the following figures. (a) A square with 30 mm sides. (b) A rectangle 40 mm by 60 mm. (c) A right-angled triangle with sides in the proportion 3:4:5. (d) An equilateral triangle with 50 mm sides. (e) A hexagon with 25 mm sides. (f) A pentagon with 30 mm sides.

4 Sketch a rectangle 40 mm by 50 mm with a 10 mm radius in each corner.

5 Sketch a square with 20 mm sides. On each side as diameter add a semicircle. Show the centre lines.

6 Sketch circles of (a) 20 mm diameter, (b) 30 mm diameter, (c) 60 mm diameter, (d) 75 mm diameter, (e) 100 mm diameter. Dimension the diameters and show the centre lines.

7 A special washer is in the form of a square with 50 mm sides and a 20 mm diameter hole in the centre. The washer is 10 mm thick. Make (a) a fully dimensioned sketch of the washer in orthographic projection, (b) an isometric sketch, (c) an oblique sketch.

8 A conical spacer is 70 mm long with a large-end diameter of 40 mm and a small-end diameter of 30 mm. Axially through the centre a 20 mm diameter hole is drilled. Make a fully dimensioned orthographic sketch of the spacer with one view as a full section. Also make isometric and oblique sketches of the component.

9 A component made from 10 mm thick plate is rectangular with sides of 35 mm and 65 mm. The corners of the rectangle are removed by 10 mm chamfers at 45°. In the centre of the plate is a 20 mm diameter hole. Make an orthographic sketch of the component with dimensions, and isometric and oblique sketches.

10 A bar with a square cross-section of 40 mm side is 120 mm long. At one end a length of 50 mm is turned down to 30 mm diameter. 30 mm from the other end a 20 mm diameter hole is drilled right through the bar, the hole centre line being perpendicular to the top face. Make a dimensioned orthographic sketch of the part, and isometric and oblique sketches.

11 A casting consists of a square base surmounted by a conical boss. The sides of the base are 140 mm long and the square has 20 mm radii in the corners. At the centres of these radii four 20 mm diameter holes are drilled right through the base and spotfaced to 30 mm diameter on the top face. The base is 30 mm thick and the bottom surface is machined. The conical boss is 80 mm high with diameters of 100 mm where it runs into the base, and 70 mm at the top. The top is machined. A hole 40 mm diameter is bored right through the casting on the axis of the boss, and counterbored at the top to 55 mm diameter and 25 mm deep. Fillets and rounds are 3 mm radius.

 Make a dimensioned orthographic sketch of the casting with sufficient information for it to be made. The material is to be cast iron.

12 A shaft is to be machined from 40 mm diameter stock. The first operation is to turn a length of 90 mm to a diameter of 30 mm. The second operation is to reduce this 30 mm diameter to 20 mm diameter for a length of 35 mm. After chamfering for 2 mm at 45°, the 20 mm diameter is threaded M20–6g, the minimum length of full form thread being 28 mm. The shaft is parted off leaving a head 40 mm diameter and 12 mm long. It is then reversed in the chuck, the head faced to 10 mm long, and drilled 10 mm diameter, 60 mm deep. Two parallel flats are milled across the head, the distance between them being 32 mm. The final operation is to drill a 2 mm diameter hole for a split pin, 6 mm from the end of the thread.

 If the material is to be mild steel and five shafts are required, make a fully dimensioned orthographic sketch of the component.

17

MACHINE DRAWINGS

UNLESS otherwise stated, the solutions to the following problems are to be drawn full size. First or Third Angle projection is to be used as required by the question. Unless asked for, do not show any hidden detail.

Bolster block
A pictorial view of a bolster block is shown opposite. Draw the following views in First Angle projection:
A front elevation, viewing the part in the direction of the arrow.
A plan projected from the front elevation.
An end view positioned on the right of the front elevation.
Show all hidden detail.

Fulcrum support
Using the pictorial view of this detail shown opposite, draw the following views in Third Angle projection, showing all hidden detail:

A front elevation, viewing the detail in the direction of the arrow.
A plan projected from the front elevation.
Both end views.

Locating bracket
A pictorial view of a locating bracket is given on page 192. Draw in First Angle projection the following views, showing all hidden detail:

A front elevation, viewing the part in the direction of the arrow.
A plan projected from the front elevation.
An end view positioned on the left of the front elevation.

Spigot
This detail is shown pictorially on page 192. Draw the following views with hidden detail in Third Angle projection:

A front elevation looking in the direction of the arrow.
A plan projected from the front elevation.
An end view positioned on the right of the front elevation.

MACHINE DRAWING

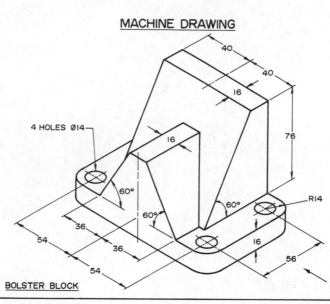

4 HOLES Ø14

40

40

16

76

16

60°

60°

60°

R14

54

36

36

16

54

56

BOLSTER BLOCK

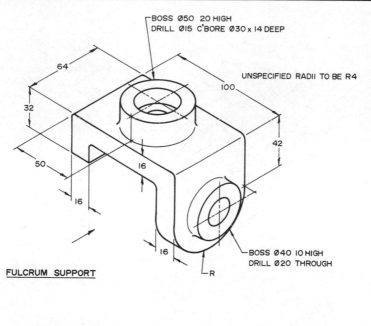

BOSS Ø50 20 HIGH
DRILL Ø15 C'BORE Ø30 x 14 DEEP

UNSPECIFIED RADII TO BE R4

64

100

32

42

50

16

16

16

BOSS Ø40 10 HIGH
DRILL Ø20 THROUGH

R

FULCRUM SUPPORT

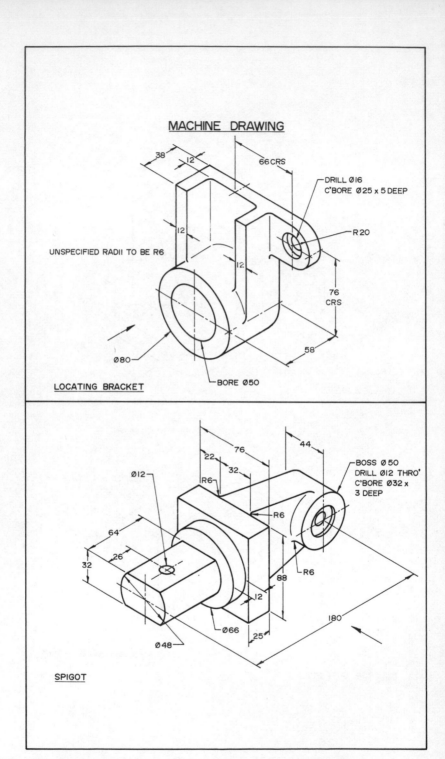

MACHINE DRAWING

DRILL Ø16
C'BORE Ø25 x 5 DEEP

R20

66 CRS

38

12

12

12

UNSPECIFIED RADII TO BE R6

76
CRS

Ø80

58

BORE Ø50

LOCATING BRACKET

76

22

32

44

BOSS Ø50
DRILL Ø12 THRO'
C'BORE Ø32 x
3 DEEP

Ø12

R6

R6

64

26

32

R6

88

12

Ø66

180

Ø48

25

SPIGOT

192

Pivot block

A pivot block is illustrated pictorially on page 194. Draw the following views in First Angle projection, with hidden detail in the plan and end view:

A sectional front elevation on AA.
A plan view.
An end view obtained by looking at the 20 mm diameter hole.

Angle bracket

This detail is shown by a pictorial view on page 194. Draw the following views in Third Angle projection, with all hidden detail:

A front elevation obtained by viewing the detail in the direction of arrow Y.
A plan view projected from the front elevation.
An end view obtained by viewing the detail in the direction of arrow X.

Bearing bracket

A plan and elevation of a bearing bracket are shown on page 195. Using First Angle projection draw the following views:

A sectional front elevation on AA.
The given plan.
An end view positioned on the right of the sectional front elevation.

Crank disc

Views of this component are shown on page 195. Draw the given front elevation and replace the given plan by a sectional plan on AA. Use Third Angle projection.

Bearing mounting

Draw the following views of the bearing mounting shown on page 196.

The given plan view.
A sectional front elevation on AA.
An end view drawn on the right of the front elevation.

Use First Angle projection.

Sealing cap

A front elevation and plan of this component are given on page 196 Draw the following views in Third Angle projection:

A half plan in place of the given plan.
A half sectional front elevation on AA.
A half sectional end view on BB.

MACHINE DRAWING

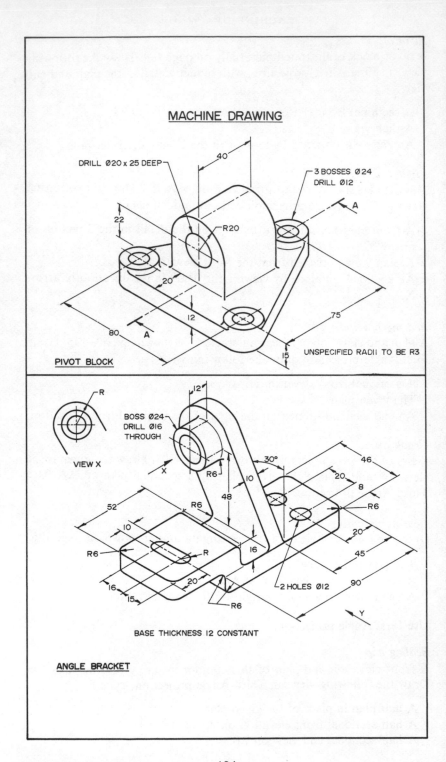

DRILL Ø20 x 25 DEEP

40

3 BOSSES Ø24
DRILL Ø12

A

22

R20

20

12

75

80

A

15

UNSPECIFIED RADII TO BE R3

PIVOT BLOCK

R

VIEW X

BOSS Ø24
DRILL Ø16
THROUGH

12

X

R6

30°

46

20

8

10

48

R6

R6

52

10

16

R6

R

16

20

45

90

2 HOLES Ø12

R6

Y

BASE THICKNESS 12 CONSTANT

ANGLE BRACKET

194

MACHINE DRAWING

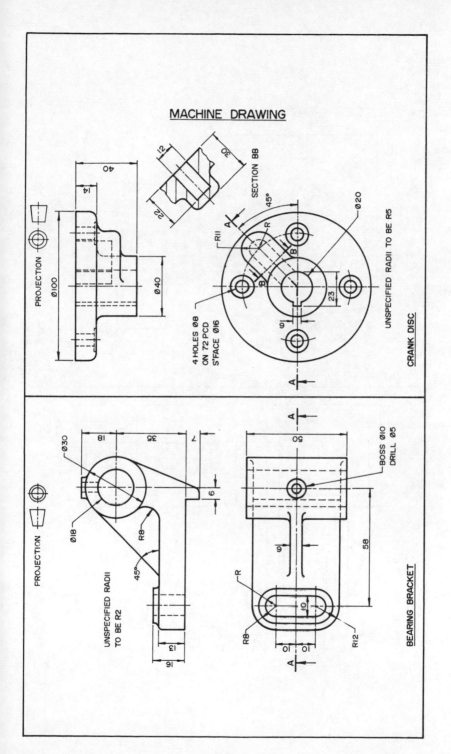

195

MACHINE DRAWING

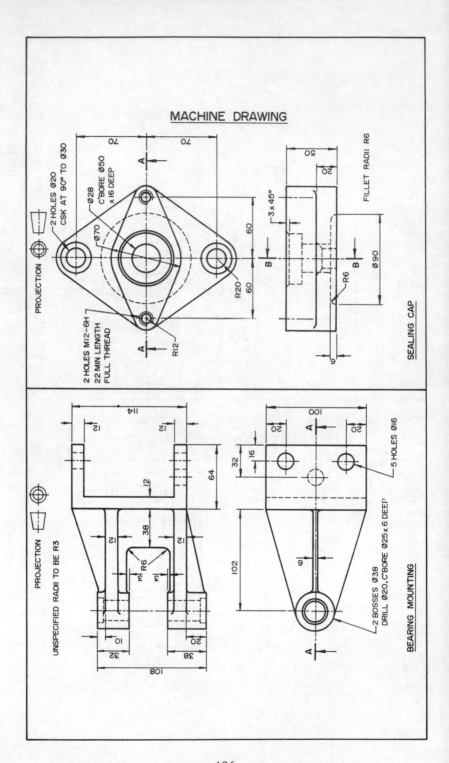

PROJECTION

2 HOLES Ø20
CSK AT 90° TO Ø30

Ø28 C'BORE Ø50
x16 DEEP

Ø70

70

70

60

60

R20

R12

2 HOLES M12-6H
22 MIN LENGTH
FULL THREAD

A

A

B

B

3 x 45°

R6

9

Ø90

50

20

FILLET RADII R6

SEALING CAP

PROJECTION

UNSPECIFIED RADII TO BE R3

114

12

12

64

12

38

12

R6

3

3

10

20

32

38

108

100

20

20

32

16

102

9

A

A

5 HOLES Ø16

2 BOSSES Ø38
DRILL Ø20, C'BORE Ø25 x 6 DEEP

BEARING MOUNTING

196

Clutch bracket

Views of a clutch bracket are shown on page 198. Draw in First Angle projection, the following views of the detail:

A sectional front elevation on BB.
A sectional end view on AA.
The given plan view.

Spindle housing

This detail is illustrated on page 198. Draw the given end view, a new sectional front elevation on AA, and a plan projected from this front elevation. Use Third Angle projection.

Angle bracket

A front elevation and plan of an angle bracket are shown on page 199. Using First Angle projection draw the given plan and from it project a new sectional front elevation on AA. Add a sectional end view on BB.

Intermediate casing

A plan and front elevation of this detail are given on page 199. Do not draw these views but replace them by a sectional front elevation on AA, a sectional plan on BB, and an end view looking on the 20 mm diameter boss.

Column stand

On page 200 are shown a front elevation and a half plan in section of a column stand. Using Third Angle projection draw the following views of this component:

A front elevation in section on AA.
An outside half plan.
An end view looking in the direction of arrow X.

Clamp assembly

The details for a small clamp assembly for a handrail are given on page 200. Draw the following views in First Angle projection with the parts assembled:

A half sectional front elevation on BB.
A plan view.
A sectional end view on AA.

To complete this assembly note that two M12 hexagon head bolts are required. The length of these bolts is to be decided by the student.

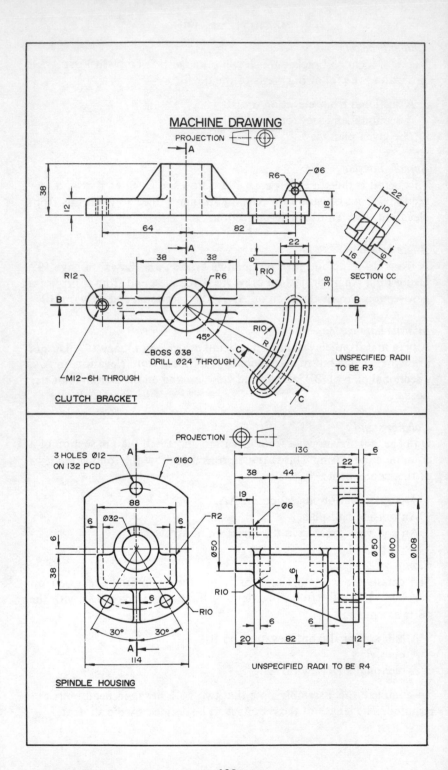

MACHINE DRAWING

PROJECTION

CLUTCH BRACKET

SECTION CC

BOSS Ø38
DRILL Ø24 THROUGH

M12-6H THROUGH

R12

R6

R10

R10

R

45°

UNSPECIFIED RADII
TO BE R3

R6   Ø6

SPINDLE HOUSING

PROJECTION

3 HOLES Ø12
ON 132 PCD

Ø160

Ø32

R2

R10

30°   30°

Ø6

R10

Ø50

Ø50

Ø100

Ø108

UNSPECIFIED RADII TO BE R4

198

MACHINE DRAWING

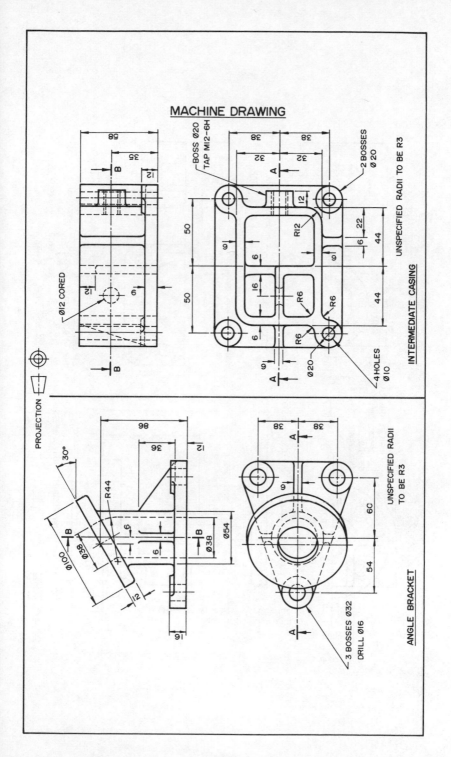

INTERMEDIATE CASING

ANGLE BRACKET

MACHINE DRAWING

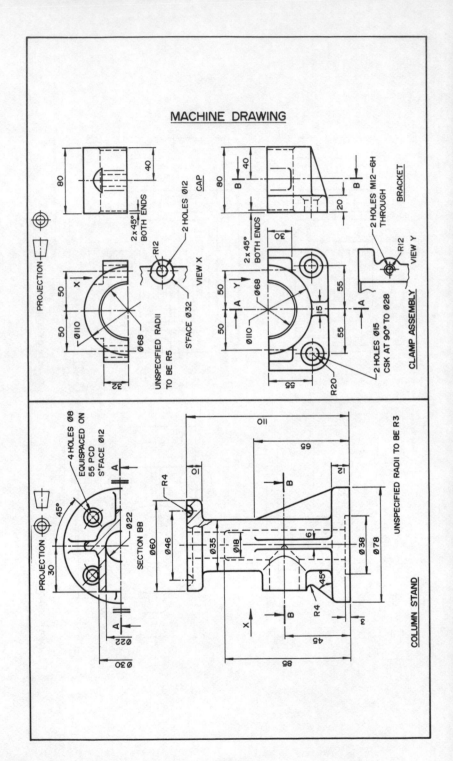

200

Indicator sleeve

The drawing on page 202 shows a front elevation and plan of this component. Do not draw the given views. Instead, draw a new front elevation obtained by viewing the given front elevation in the direction of arrow A. From this view project a sectional plan on BB and a sectional end view. The section plane for the end view is to pass through the centre line of the component and the view is to be drawn on the right-hand side of the front elevation. Use First Angle projection.

Tappet lever

Draw in First Angle projection the following views of the tappet lever shown on page 203:

The given front elevation.
A sectional plan on BB.
A sectional end view on AA.

Bearing

The drawing on page 203 shows a front elevation and end view of a special bearing. Using Third Angle projection draw the given front elevation and project from it a sectional auxiliary plan on AA.

Junction box

A front elevation and two end views of this detail are shown on page 204. Draw the given front elevation and project from it a sectional view on AA. Use Third Angle projection.

Housing fixture bracket

Views of this bracket are shown on page 205. Draw the given end view and from it project a new sectional front elevation on AA. From this front elevation project a plan. Use Third Angle projection.

Double mounting bracket

Draw the following views of this detail, which is shown on page 206. Use First Angle projection:

The given plan view.
A sectional front elevation on AA.
A sectional end view on BB.
A part section on CC.

Machine vice base

A plan, front elevation and end view of a machine vice base are shown on page 207. Draw in First Angle projection the given plan, a sectional front elevation on AA and a sectional end view on BB. Add a local section on the plan view through one of the M8–6H tappings.

MACHINE DRAWING

PROJECTION

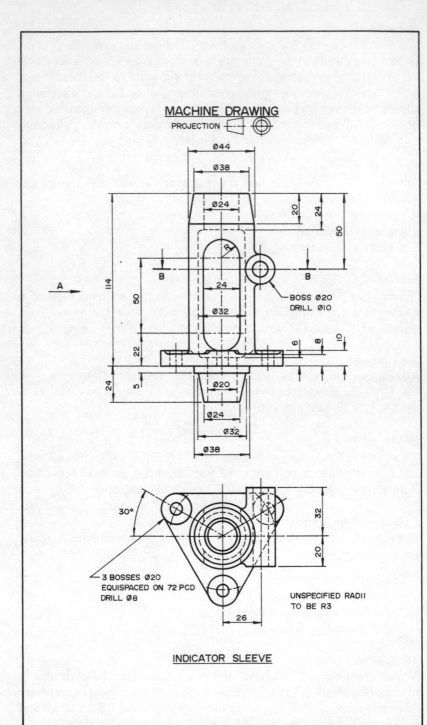

BOSS Ø20
DRILL Ø10

3 BOSSES Ø20
EQUISPACED ON 72 PCD
DRILL Ø8

UNSPECIFIED RADII
TO BE R3

INDICATOR SLEEVE

202

MACHINE DRAWING

PROJECTION

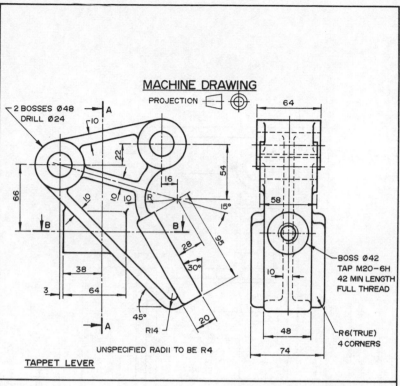

2 BOSSES Ø48 DRILL Ø24

22

16

R

15°

54

10

66

10 10

B

B

28

30°

95

38

64

3

45°

R14

20

UNSPECIFIED RADII TO BE R4

TAPPET LEVER

64

58

10

BOSS Ø42
TAP M20-6H
42 MIN LENGTH
FULL THREAD

48

74

R6(TRUE)
4 CORNERS

PROJECTION

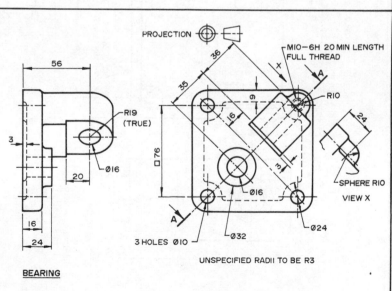

56

R19
(TRUE)

3

20

Ø16

16

24

36

35

M10-6H 20 MIN LENGTH
FULL THREAD

X

A

R10

6

16

□76

3

24

A

Ø16

Ø32

3 HOLES Ø10

Ø24

SPHERE R10

VIEW X

UNSPECIFIED RADII TO BE R3

BEARING

203

MACHINE DRAWING

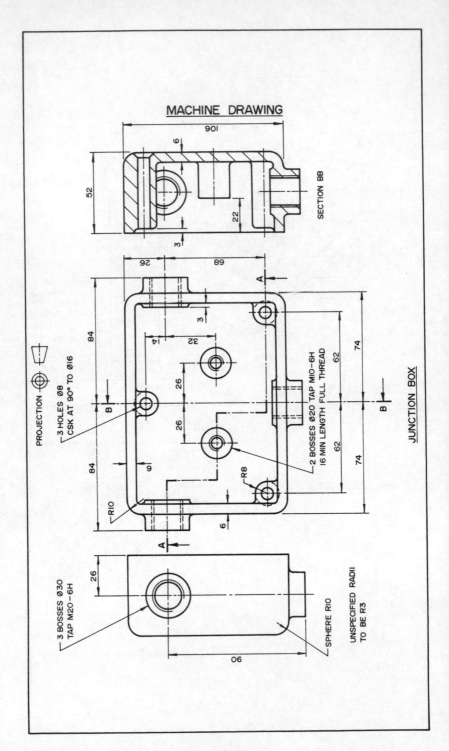

JUNCTION BOX

SECTION BB

PROJECTION

901

52

6

3

22

26

68

84

3

32

14

26

26

62

62

74

74

B

B

A

A

9

RIO

R8

3 HOLES Ø8
CSK AT 90° TO Ø16

2 BOSSES Ø20 TAP M10-6H
16 MIN LENGTH FULL THREAD

84

6

3 BOSSES Ø30
TAP M20-6H

26

06

SPHERE RIO

UNSPECIFIED RADII
TO BE R3

204

MACHINE DRAWING

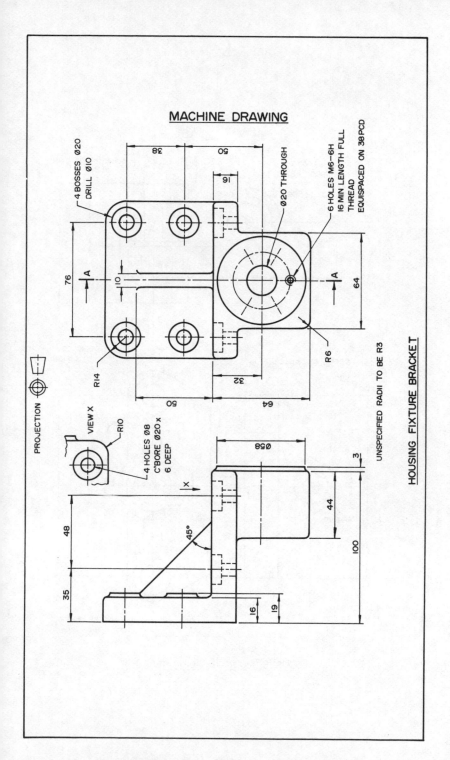

PROJECTION

HOUSING FIXTURE BRACKET

UNSPECIFIED RADII TO BE R3

4 BOSSES Ø20
DRILL Ø10

Ø20 THROUGH

6 HOLES M6-6H
16 MIN LENGTH FULL
THREAD
EQUISPACED ON 38PCD

38
50
16
76
110
64
A
A
R14
R6
50
32
64

VIEW X
R10

4 HOLES Ø8
C'BORE Ø20 x
6 DEEP
X

Ø58
48
35
100
44
3
45°
16
19

205

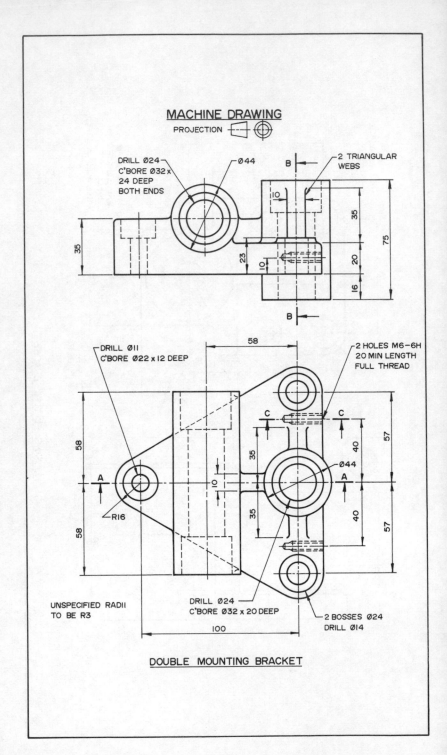

MACHINE DRAWING

PROJECTION

DRILL Ø24
C'BORE Ø32 x
24 DEEP
BOTH ENDS

Ø44

B

2 TRIANGULAR
WEBS

10

35

75

23

20

10

16

B

DRILL Ø11
C'BORE Ø22 x 12 DEEP

58

2 HOLES M6-6H
20 MIN LENGTH
FULL THREAD

58

C

C

57

35

10

40

A

Ø44

A

35

40

57

R16

58

UNSPECIFIED RADII
TO BE R3

DRILL Ø24
C'BORE Ø32 x 20 DEEP

2 BOSSES Ø24
DRILL Ø14

100

DOUBLE MOUNTING BRACKET

MACHINE DRAWING

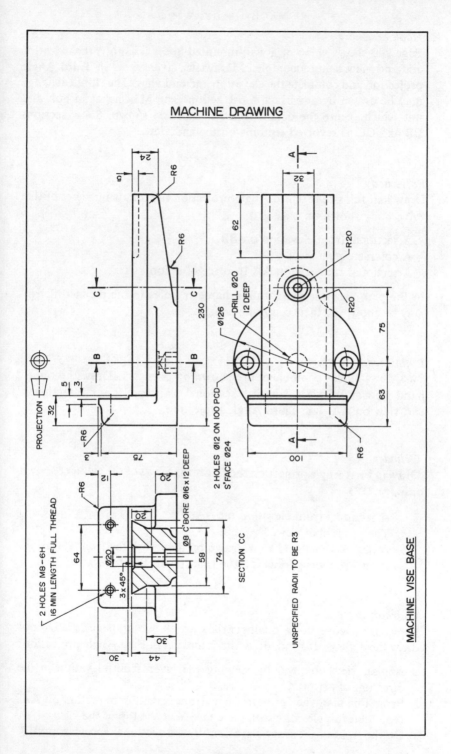

MACHINE VISE BASE

PROJECTION

SECTION CC

UNSPECIFIED RADII TO BE R3

2 HOLES M8-6H
16 MIN LENGTH FULL THREAD

Ø8 C'BORE Ø16×12 DEEP

3×45°

2 HOLES Ø12 ON 100 PCD
S'FACE Ø24

DRILL Ø20
12 DEEP

Ø126

R6

R20

R20

Hook assembly

Page 209 shows views of a wall-mounted hook assembly, the elevation and end view being incomplete. Draw the given views in Third Angle projection and complete the elevation and end view. The elliptical flange may be drawn by any exact construction. The M12 hexagon bolt and nut which secure the details together must be shown. Show sections BB and CC as revolved sections on the end view.

Valve body

Draw half full size in Third Angle projection the following views of the valve body shown on page 210:

A sectional front elevation on BB.
A complete plan view.
A sectional end view on AA in place of section CC.

In the plan view the two 20 mm diameter holes in the inclined flange may be shown by their centre lines only.

Lathe steady casting

Two views of this component are shown on page 211. Draw the right-hand view and project from it a sectional plan on BB and a sectional end view on AA. Use Third Angle projection.

Cylinder

Draw in First Angle projection the following views of the cylinder shown on page 212:

A sectional front elevation on AA.
The given plan view.
A half sectional end view drawn on the left of the front elevation.
A half end view drawn on the right of the front elevation.

Gearbox cover

Views of the cover for a small gearbox are shown on page 213. Do not draw these views. Instead, draw the following in First Angle projection:

A plan view obtained by viewing the given front elevation in the direction of arrow X.
From this plan view project a new front elevation in section on AA. Note that this elevation will have face Y at the top of the view.
On the right of the front elevation draw a sectional end view on BB.

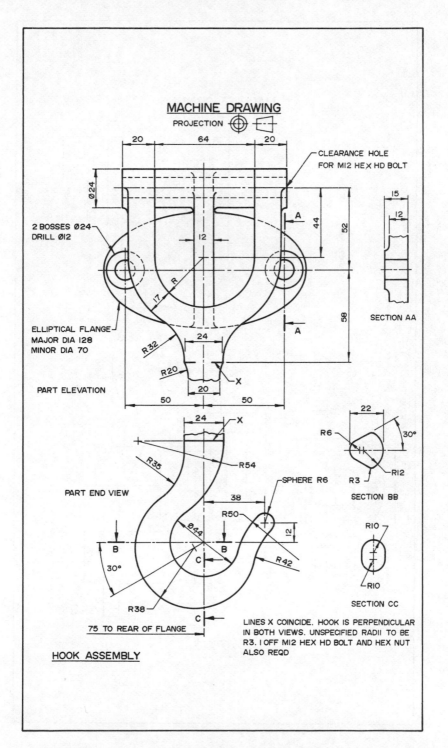

MACHINE DRAWING

PROJECTION

20 | 64 | 20

CLEARANCE HOLE
FOR MI2 HEX HD BOLT

Ø24

A

44

52

2 BOSSES Ø24
DRILL Ø12

15

12

12

R

17

SECTION AA

ELLIPTICAL FLANGE
MAJOR DIA 128
MINOR DIA 70

58

R32

24

R20

X

PART ELEVATION

20

50 | 50

24 — X

PART END VIEW

R35

R54

22

R6

30°

SPHERE R6

R12

R3

38

R50

12

SECTION BB

Ø44

B | B

30°

C

C

R42

R10

R38

R10

C

75 TO REAR OF FLANGE

SECTION CC

LINES X COINCIDE. HOOK IS PERPENDICULAR
IN BOTH VIEWS. UNSPECIFIED RADII TO BE
R3. I OFF MI2 HEX HD BOLT AND HEX NUT
ALSO REQD

HOOK ASSEMBLY

209

MACHINE DRAWING

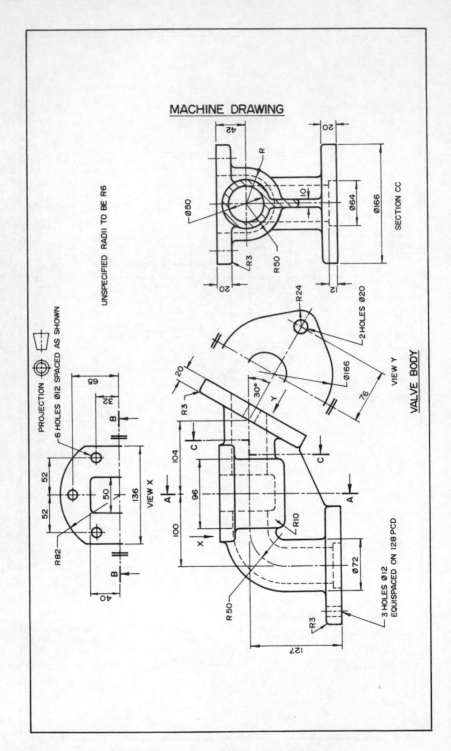

UNSPECIFIED RADII TO BE R6

PROJECTION ⊕

6 HOLES Ø12 SPACED AS SHOWN

VIEW X

VIEW Y

SECTION CC

VALVE BODY

2 HOLES Ø20

3 HOLES Ø12
EQUISPACED ON 128 PCD

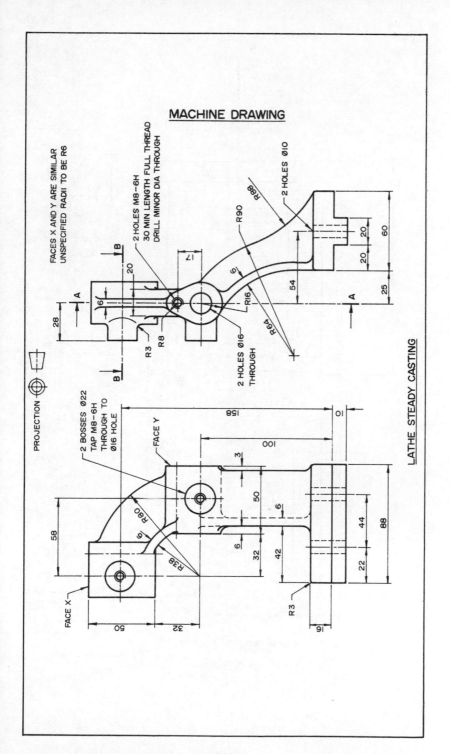

MACHINE DRAWING

FACES X AND Y ARE SIMILAR
UNSPECIFIED RADII TO BE R6

2 HOLES M8-6H
30 MIN LENGTH FULL THREAD
DRILL MINOR DIA THROUGH

2 HOLES Ø10

R88

R90

R16

R64

2 HOLES Ø16
THROUGH

R3   R8

6

9

20

28

6

17

20

20

60

54

25

B

A

A

B

PROJECTION

2 BOSSES Ø22
TAP M8-6H
THROUGH TO
Ø16 HOLE

FACE Y

FACE X

R80

R38

6

58

50

32

158

100

3

50

6

6

32

42

88

44

22

10

16

R3

LATHE STEADY CASTING

211

MACHINE DRAWING

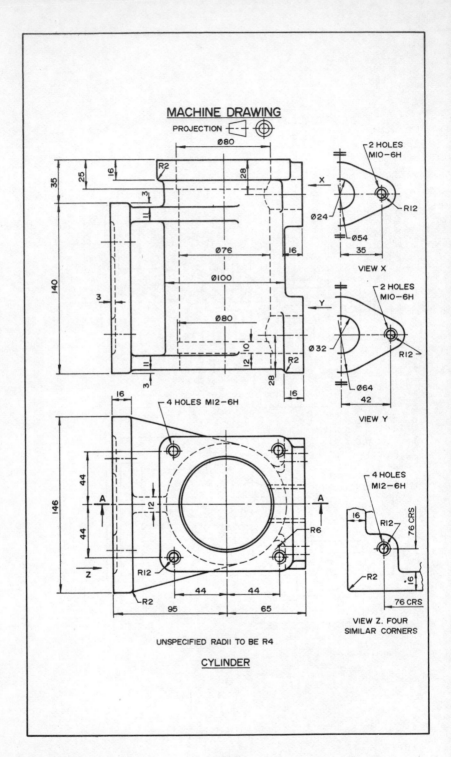

PROJECTION

Ø80

R2

35 25 16 28

3

Ø76 16

140 Ø100

3 Ø80

10 12 R2

3 28 16

16 4 HOLES M12−6H

44

A 12 A

146

44

Z R6

R12 44 44 R6

R2

95 65

R2

UNSPECIFIED RADII TO BE R4

CYLINDER

2 HOLES
M10−6H

Ø24 R12

Ø54

35

VIEW X

2 HOLES
M10−6H

Ø32 R12

Ø64

42

VIEW Y

4 HOLES
M12−6H

16 R12 76 CRS

R2 16

76 CRS

VIEW Z. FOUR SIMILAR CORNERS

MACHINE DRAWING

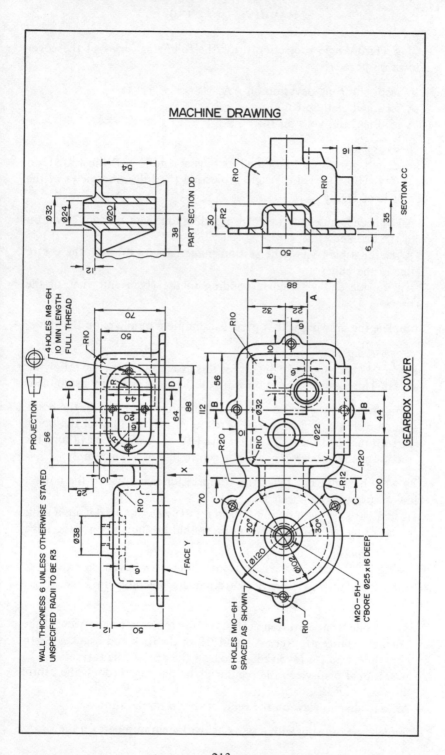

WALL THICKNESS 6 UNLESS OTHERWISE STATED
UNSPECIFIED RADII TO BE R3

PART SECTION DD

SECTION CC

GEARBOX COVER

PROJECTION

4 HOLES M8-6H
10 MIN LENGTH
FULL THREAD

FACE Y

6 HOLES M10-6H
SPACED AS SHOWN

M20-5H
C'BORE Ø25×16 DEEP

213

Cover
Using Third Angle projection draw the following views of the cover shown on page 215:

A sectional front elevation on AA.
A sectional plan on CC.
A sectional end view on BB.

Pulley assembly
The details for a wall-mounted pulley assembly are illustrated on page 216. Draw in Third Angle projection the following views of the complete assembly:

An outside front elevation corresponding to the given front elevation of the bracket.
A sectional plan view, the section plane passing through the centre line of the pulley.
An outside end view corresponding to the given end view of the bracket.

Complete the drawing with a parts list and item numbers on the views.

Cylinder relief valve
On page 217 are shown the details for this assembly. Draw the following views in Third Angle projection, with all the parts correctly assembled:

An outside front elevation corresponding to section AA of the body.
A complete plan, corresponding to the given half plan of the body.
A sectional end view drawn on the left of the front elevation, the section plane passing through the centre line of the assembly.

The M20 hexagon nut locks the compression screw when the correct blow-off pressure has been set.
Add a parts list with item numbers on the views and draw up a table for the spring particulars. The spring should be shown conventionally.

Strut attachment
Drawings of the details for a strut attachment are given on page 218. Draw the following views of the complete assembly in First Angle projection:

An outside front elevation corresponding to the left-hand view of the bracket. The centre lines AA and BB of the fork and bracket are to be in line. Show a revolved section on the arm of the fork.
A sectional plan view, the section plane passing through the centre lines AA and BB.
An outside end view on the right of the front elevation.

Complete the drawing with a parts list with item numbers on the views.

MACHINE DRAWING

PROJECTION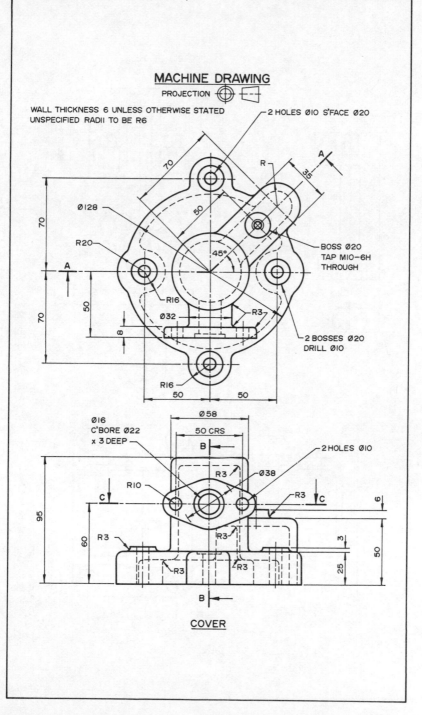

WALL THICKNESS 6 UNLESS OTHERWISE STATED
UNSPECIFIED RADII TO BE R6

2 HOLES Ø10 S'FACE Ø20

70

R

35

A

Ø128

50

A

R20

45°

BOSS Ø20
TAP M10-6H
THROUGH

70

70

R16

Ø32

R3

50

8

2 BOSSES Ø20
DRILL Ø10

R16

50   50

Ø58

Ø16
C'BORE Ø22
x 3 DEEP

50 CRS

B

2 HOLES Ø10

R3

Ø38

R10

R3

C   C

6

95

60

R3

R3

3

R3   R3

50

25

B

COVER

215

MACHINE DRAWING

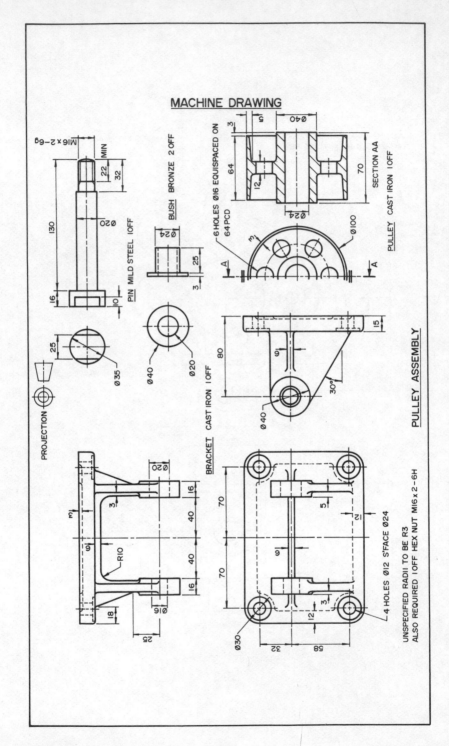

MI6 × 2-6g

22 MIN

32

130

Ø20

16

10

25

Ø35

PROJECTION

BUSH BRONZE 2 OFF

Ø24

25

3

Ø40

Ø20

PIN MILD STEEL 1 OFF

6 HOLES Ø16 EQUISPACED ON 64 PCD

3

5

Ø40

64

12

70

Ø24

SECTION AA

PULLEY CAST IRON 1 OFF

Ø100

A

A

BRACKET CAST IRON 1 OFF

15

80

6

30°

Ø40

PULLEY ASSEMBLY

Ø20

16

3

40

40

16

6

R10

18

25

Ø16

70

5

12

6

12

3

70

Ø30

32

58

4 HOLES Ø12 S'FACE Ø24

UNSPECIFIED RADII TO BE R3

ALSO REQUIRED 1 OFF HEX NUT MI6 × 2 - 6H

216

MACHINE DRAWING

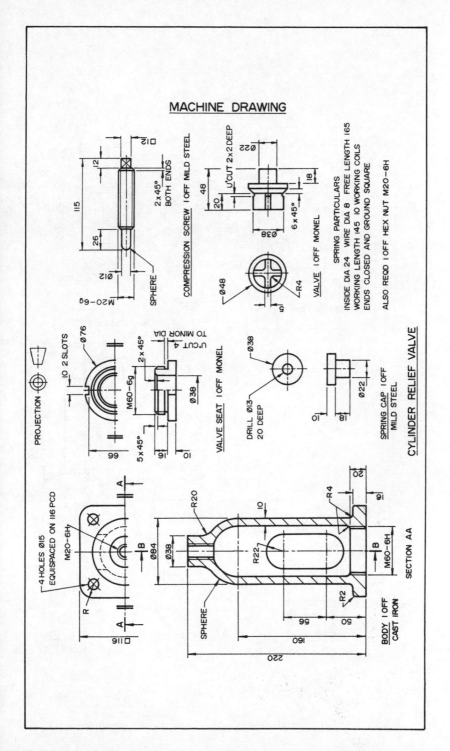

COMPRESSION SCREW 1 OFF MILD STEEL

VALVE 1 OFF MONEL

SPRING PARTICULARS

INSIDE DIA 24 WIRE DIA 8 FREE LENGTH 165
WORKING LENGTH 145 10 WORKING COILS
ENDS CLOSED AND GROUND SQUARE

ALSO REQD 1 OFF HEX NUT M20-6H

VALVE SEAT 1 OFF MONEL

SPRING CAP 1 OFF
MILD STEEL

CYLINDER RELIEF VALVE

BODY 1 OFF
CAST IRON

SECTION AA

217

MACHINE DRAWING

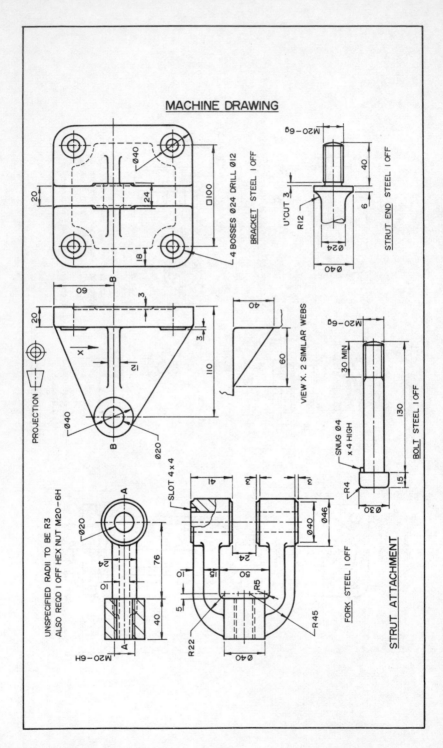

Ø40

□100

20

24

18

4 BOSSES Ø24 DRILL Ø12

BRACKET STEEL I OFF

M20-6g

40

U'CUT 3
R12

6

Ø24

Ø40

STRUT END STEEL I OFF

60

3

20

3

110

X

12

Ø40

Ø20

B

B

PROJECTION

40

60

VIEW X. 2 SIMILAR WEBS

M20-6g

30 MIN

130

SNUG Ø4
x 4 HIGH

R4

Ø30

15

BOLT STEEL I OFF

UNSPECIFIED RADII TO BE R3
ALSO REQD I OFF HEX NUT M20-6H

A

Ø20

76

24

10

40

A

M20-6H

SLOT 4 x 4

14

3

3

Ø46

Ø40

24

10

15

50

5

R5

R22

R45

Ø40

FORK STEEL I OFF

STRUT ATTACHMENT

218

Flange coupling

Page 220 shows the details for a flange coupling. Draw, in Third Angle projection, the following views of the assembly with shaft ends and keys in position:

An outside elevation corresponding to the given part elevation of the flange.

A sectional end view on AA. On this view show a broken-out section on one shaft around the key.

Key and keyway dimensions have been omitted on page 220. Select a suitable key from the table on page 223.

The driving pins are attached to each flange alternately by a nut and washer. The bushes are centrally placed in the 28 mm diameter bores and retained on the driving pins by a nut and washer. Note that the parts are assembled so that there is a gap between the flanges and shaft ends.

Complete the drawing with a parts list and item numbers on the views.

MACHINE DRAWING

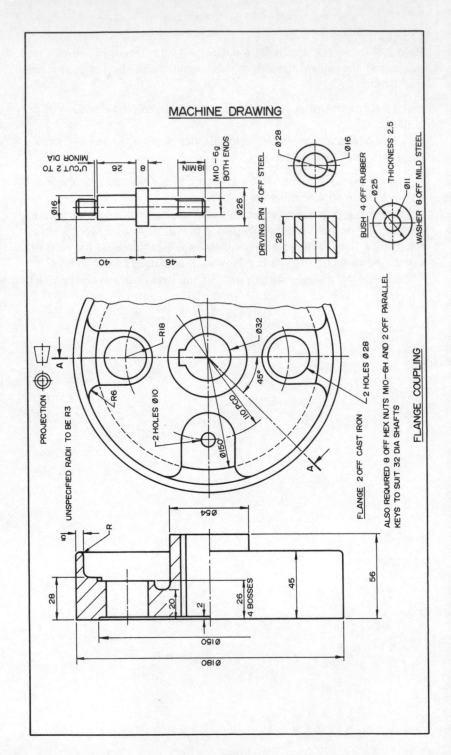

PROJECTION

UNSPECIFIED RADII TO BE R3

DRIVING PIN 4 OFF STEEL

U'CUT 2 TO MINOR DIA

M10 - 6g BOTH ENDS

Ø16

Ø26

26   8   18 MIN

40   46

BUSH 4 OFF RUBBER

Ø28

Ø16

28

WASHER 8 OFF MILD STEEL

Ø25

Ø11

THICKNESS 2.5

FLANGE 2 OFF CAST IRON

R18

R6

Ø32

45°

110 PCD

2 HOLES Ø10

2 HOLES Ø28

Ø150

ALSO REQUIRED 8 OFF HEX NUTS M10-6H AND 2 OFF PARALLEL
KEYS TO SUIT 32 DIA SHAFTS

FLANGE COUPLING

Ø54

Ø150

Ø180

28

20

26   4 BOSSES

2

45

56

R

220

DRAWING SHEET SIZES

Designation	Size, millimetres
A0	841 x 1189
A1	594 x 841
A2	420 x 594
A3	297 x 420
A4	210 x 297

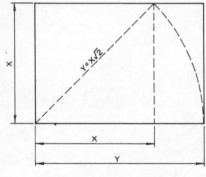

Basic size, A0, of the A series is a rectangle with an area of 1 square metre (X x Y = 1). The sides of the rectangle satisfy the proportion

$$X:Y::1:\sqrt{2}$$

Smaller sizes are obtained by halving larger dimensions and larger sizes by doubling the smaller dimensions. The ratio of length to breadth is the same for all sizes.
The A4 size is the same as that used for general office stationery, report forms, letter paper etc.

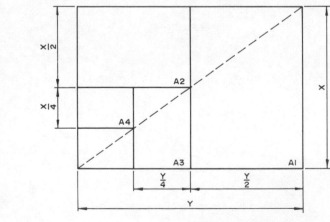

222

ISO METRIC PRECISION HEXAGON BOLT, SCREW AND NUT DIMENSIONS

All dimensions in millimetres

NOMINAL SIZE	PITCH OF THREAD		WIDTH ACROSS FLATS (MAX)	WASHER FACE DIA (MAX)	WASHER FACE DEPTH	HEIGHT OF HEAD (NOM'L)	NUT THICKNESS (MAX)	
	COARSE	FINE					NORMAL	THIN
M1.6	0.35	–	3.2	–	–	1.10	1.30	–
M2	0.40	–	4.0	–	–	1.40	1.60	–
M2.5	0.45	–	5.0	–	–	1.70	2.00	–
M3	0.50	–	5.5	6.08	0.1	2.00	2.40	–
M4	0.70	–	7.0	6.55	0.1	2.80	3.20	–
M5	0.80	–	8.0	7.55	0.2	3.50	4.00	–
M6	1.00	–	10.0	9.48	0.3	4.00	5.00	–
M8	1.25	1.00	13.0	12.43	0.4	5.50	6.50	5.0
M10	1.50	1.25	17.0	16.43	0.4	7.00	8.00	6.0
M12	1.75	1.25	19.0	18.37	0.4	8.00	10.00	7.0
(M14)	2.00	1.50	22.0	21.37	0.4	9.00	11.00	8.0
M16	2.00	1.50	24.0	23.27	0.4	10.00	13.00	8.0
(M18)	2.50	1.50	27.0	26.27	0.4	12.00	15.00	9.0
M20	2.50	1.50	30.0	29.27	0.4	13.00	16.00	9.0
(M22)	2.50	1.50	32.0	31.21	0.4	14.00	18.00	10.0
M24	3.00	2.00	36.0	34.98	0.5	15.00	19.00	10.0
(M27)	3.00	2.00	41.0	39.98	0.5	17.00	22.00	12.0
M30	3.50	2.00	46.0	44.98	0.5	19.00	24.00	12.0
(M33)	3.50	2.00	50.0	48.98	0.5	21.00	26.00	14.0
M36	4.00	3.00	55.0	53.86	0.5	23.00	29.00	14.0
(M39)	4.00	3.00	60.0	58.86	0.6	25.00	31.00	16.0
M42	4.50	–	65.0	63.76	0.6	26.00	34.00	16.0
(M45)	4.50	–	70.0	68.76	0.6	28.00	36.00	18.0
M48	5.00	–	75.0	73.76	0.6	30.00	38.00	18.0
(M52)	5.00	–	80.0	–	–	33.00	42.00	20.0
M56	5.50	–	85.0	–	–	35.00	45.00	–
(M60)	5.50	–	90.0	–	–	38.00	48.00	–
M64	6.00	–	95.0	–	–	40.00	51.00	–
(M68)	6.00	–	100.0	–	–	43.00	54.00	–

Sizes in brackets are non—preferred

Based on BS 3692:1967. Fine pitch series from BS 3643: Part I:1963

KEY AND KEYWAY DIMENSIONS FROM BS 4235:Part 1:1972

SQUARE AND RECTANGULAR PARALLEL KEYS

All dimensions in millimetres

SHAFT		KEY				KEYWAY			
NOMINAL DIA d		SECTION b x h WIDTH x THICKNESS	CHAMFER s MAX	RANGE OF LENGTHS l		WIDTH b NOM	DEPTH		RAD r MIN
							SHAFT t_1 NOM	HUB t_2 NOM	
OVER	INCL			OVER	INCL				
6	8	2 x 2	0.25	6	20	2	1.2	1	0.08
8	10	3 x 3	0.25	6	36	3	1.8	1.4	0.08
10	12	4 x 4	0.25	8	45	4	2.5	1.8	0.08
12	17	5 x 5	0.40	10	56	5	3	2.3	0.16
17	22	6 x 6	0.40	14	70	6	3.5	2.8	0.16
22	30	8 x 7	0.40	18	90	8	4	3.3	0.16
30	38	10 x 8	0.60	22	110	10	5	3.3	0.25
38	44	12 x 8	0.60	28	140	12	5	3.3	0.25
44	50	14 x 9	0.60	36	160	14	5.5	3.8	0.25
50	58	16 x 10	0.60	45	180	16	6	4.3	0.25
58	65	18 x 11	0.60	50	200	18	7	4.4	0.25
65	75	20 x 12	0.80	56	220	20	7.5	4.9	0.40

SQUARE AND RECTANGULAR TAPER KEYS

SHAFT		KEY					KEYWAY			
NOMINAL DIA d		SECTION b x h WIDTH x THICKNESS	CHAMFER s MAX	RANGE OF LENGTHS l		GIB HEAD h_1 NOM	WIDTH b SHAFT & HUB NOM	DEPTH		RAD r MIN
								SHAFT t_1 NOM	HUB t_2 NOM	
OVER	INCL			OVER	INCL					
6	8	2 x 2	0.25	6	20	—	2	1.2	0.5	0.08
8	10	3 x 3	0.25	6	36	—	3	1.8	0.9	0.08
10	12	4 x 4	0.25	8	45	7	4	2.5	1.2	0.08
12	17	5 x 5	0.40	10	56	8	5	3	1.7	0.16
17	22	6 x 6	0.40	14	70	10	6	3.5	2.2	0.16
22	30	8 x 7	0.40	18	90	11	8	4	2.4	0.16
30	38	10 x 8	0.60	22	110	12	10	5	2.4	0.25
38	44	12 x 8	0.60	28	140	12	12	5	2.4	0.25
44	50	14 x 9	0.60	36	160	14	14	5.5	2.9	0.25
50	58	16 x 10	0.60	45	180	16	16	6	3.4	0.25
58	65	18 x 11	0.60	50	200	18	18	7	3.4	0.25
65	75	20 x 12	0.80	56	220	20	20	7.5	3.9	0.40

Diameter ranges extend to 500mm

BRITISH STANDARD PARALLEL PIPE THREAD DIMENSIONS

NOMINAL SIZE	THREADS PER INCH	PITCH	DEPTH OF THREAD	MAJOR DIAMETER	PITCH DIAMETER	MINOR DIAMETER
		mm	mm	mm	mm	mm
1/16	28	0.907	0.581	7.723	7.142	6.561
1/8	28	0.907	0.581	9.728	9.147	8.566
1/4	19	1.337	0.856	13.157	12.301	11.445
3/8	19	1.337	0.856	16.662	15.806	14.950
1/2	14	1.814	1.162	20.955	19.793	18.631
(5/8)	14	1.814	1.162	22.911	21.749	20.587
3/4	14	1.814	1.162	26.441	25.279	24.117
(7/8)	14	1.814	1.162	30.201	29.039	27.877
1	11	2.309	1.479	33.249	31.770	30.291
(1 1/8)	11	2.309	1.479	37.897	36.418	34.939
1 1/4	11	2.309	1.479	41.910	40.431	38.952
1 1/2	11	2.309	1.479	47.803	46.324	44.845
(1 3/4)	11	2.309	1.479	53.746	52.267	50.788
2	11	2.309	1.479	59.614	58.135	56.656
(2 1/4)	11	2.309	1.479	65.710	64.231	62.752
2 1/2	11	2.309	1.479	75.184	73.705	72.226
(2 3/4)	11	2.309	1.479	81.534	80.055	78.576
3	11	2.309	1.479	87.884	84.405	84.926
(3 1/2)	11	2.309	1.479	100.330	98.851	97.372
4	11	2.309	1.479	113.030	111.551	110.072
(4 1/2)	11	2.309	1.479	125.730	124.251	122.772
5	11	2.309	1.479	138.430	136.951	135.472
(5 1/2)	11	2.309	1.479	151.130	149.651	148.172
6	11	2.309	1.479	163.830	162.351	160.372

Sizes in brackets are non-preferred

Based on BS 2779 : 1973

Major diameters above are gauge diameters for internal and external
taper pipe threads. The gauge diameter lies at the gauge plane which
is theoretically located at the face of an internal thread and at the
gauge length from the small end of an external thread. See BS 21:1973

ABBREVIATIONS OF GENERAL ENGINEERING TERMS
FOR USE ON DRAWINGS

Note 1 Abbreviations are the same in the singular and plural.
2 Full stops are not used except when the abbreviation
makes a word which may be confusing, e.g. the
abbreviation for the word 'number.'

Across flats	A/F
Assembly	ASSY
Centres	CRS
Centre line	℄ or CL
Chamfered	CHAM
Cheese head	CH HD
Countersunk	CSK
Counterbore	C'BORE
Diameter (in a note)	DIA
Diameter (preceding a dimension)	∅
Drawing	DRG
External	EXT
Figure	FIG.
Hexagon	HEX
Internal	INT
Left hand	LH
Long	LG
Material	MATL
Maximum	MAX
Minimum	MIN
Number	NO.
Pitch circle diameter	PCD
Radius (preceding a dimension, capital letter only)	R
Required	REQD
Right hand	RH
Round head	RD HD
Screwed	SCR
Sheet	SH
Specification	SPEC
Spherical diameter (preceding a dimension)	SPHERE ∅
Spherical radius (preceding a dimension)	SPHERE R
Spotface	S'FACE
Square (in a note)	SQ
Square (preceding a dimension)	□
Standard	STD
Undercut	U'CUT
Taper, on diameter or width	

Based on BS 308 : Part I : 1972